A Good Start in Life

ALSO BY NORBERT HERSCHKOWITZ

The Network Brain: Its Lifelong Development

ALSO BY ELINORE CHAPMAN HERSCHKOWITZ

Feeling at Home in Bern

A Good Start in Life

Understanding Your Child's Brain and Behavior

Norbert Herschkowitz, M.D.

Elinore Chapman Herschkowitz

Joseph Henry Press
Washington, D.C.

Joseph Henry Press • 2101 Constitution Avenue, N.W. • Washington, D.C. 20418

The Joseph Henry Press, an imprint of the National Academy Press, was created with the goal of making books on science, technology, and health more widely available to professionals and the public. Joseph Henry was one of the founders of the National Academy of Sciences and a leader in early American science.

The Dana Press, a division of The Charles A. Dana Foundation, publishes health and popular science books about the brain for the general reader. The Dana Foundation is a private philanthropic organization with particular interests in health and education.

Any opinions, findings, conclusions, or recommendations expressed in this volume are those of the authors and do not necessarily reflect the views of the National Academy of Sciences or its affiliated institutions.

Library of Congress Cataloging-in-Publication Data

Herschkowitz, Norbert.
  A good start in life : understanding your child's brain and behavior / Norbert Herschkowitz, Elinore Chapman Herschkowitz ; foreword by Jerome Kagan.
     p. cm.
Includes bibliographical references and index.
  ISBN 0-309-07639-0
  1. Child psychology. 2. Child development. 3. Infant psychology. 4. Infants—Development. I. Herschkowitz, Elinore Chapman. II. Title.
  BF721 .H413 2002
  155.4—dc21

                                  2002002353

Cover photograph by PhotoDisc.

Printed in the United States of America.

Contents

v

Part Two   The First Year

Part Three   The Second Year

Part Four   Three to Six Years

Foreword

The extraordinary advances in our understanding of the brain, especially the biological changes that occur over the opening decade of life, have corrected John Locke's eighteenth-century metaphor for psychological development, which dominated the interpretation of human psychological growth for most of the last 200 years. When I was a graduate student in the middle of the last century, scholars concerned with psychological phenomena believed that the universal milestones of development—for example, the fears of infancy, language, shame, and reasoning—were the products of contingent rewards and punishments. Even the extremely deviant phenomenon of autism was, believe it or not, interpreted as the product of an indifferent, rejecting mother. This assumption is not surprising; rather, what is surprising is that my generation regarded this claim as perfectly reasonable. Whenever an intellectual community becomes convinced of the validity of a premise that can be defended rationally and is in accord with political ideology, it is not difficult to be caught up in a dogmatic defense of what may turn out to be empirically incorrect. Remember, prior to Copernicus, some very smart people were certain that the sun revolved around the earth, that witchcraft could make one ill, and that the ashes in a cold fireplace were contained originally in the log that had burned there.

Thanks to the labor and ingenuity of so many scientists, we now recognize that infants in every cultural setting, from a Sumatran jungle to a Paris apartment, will cry at the approach of a stranger after they are six months old because of the biological timetable governing brain maturation that links, for example, sensory association areas and temporal lobe structures with the prefrontal cortex. A host of other phenomena are also yoked to the maturation of the brain.

The discoveries of neuroscientists have also forced clinicians and investigators to appreciate the influence of individual variation in brain chemistry, which can lead to variation among infants in energy level, attentiveness, and especially, fear and timidity whenever a new event occurs.

Because these ideas are relatively recent, they have not been incorporated into books and articles written to help parents understand their children. It was time for such a text to be published, and Elinore and Norbert Herschkowitz have performed this service for the parents of the many millions of infants who will be born this year. This remarkable book, *A Good Start in Life*, weaves the psychological and biological knowledge into a seamless tapestry that is informing without being arcane, technically detailed without being pedantic, and especially, concise without being flippant.

The book is marked by several innovations. The authors appreciate that cognition and emotion are not autonomous processes but part of a larger whole. They are also wary of the simple determinism that many similar books adopt. The authors understand that Nature's favorite answer is not yes or no but maybe. The discussions of language, reactions to novelty, self-awareness, and personality are unique, and I know of no other book for parents that treats the notion of temperament with such sophistication.

Parents will appreciate the suggestion that the more fruitful way to view the child's intellectual capacities is as multiple cognitive abilities, not as one general intellectual competence.

It is not unimportant that these authors wisely avoid the easy temptation to instruct parents on how to feed, bathe, and play with their infants. They understand that raising children is not like making a soufflé. Young children are robust, not fragile porcelain clocks that are broken easily. More important, parents recognize that after the second birthday, children begin to interpret their experiences symbolically, and it is these experiences, not the events as recorded by a camera, that are the most important determinants of the child's future psychological profile. Most Puritan children who were punished severely by their parents did not grow up with a serious need for professional help because they interpreted the parental chastisements as having a benevolent intention; namely, to help the child build good character. Although the authors avoid favoring or opposing any one political ideology, they do take one important position that theory would regard as reasonable: they believe that it is important for parents to communicate to the child that he or she is a member of a family, a member of a team with both privileges and responsibilities. The increased affluence of so many middle-class parents has made unnecessary the chores that most nineteenth-century children were expected to perform. The fact that clothes were washed or firewood cut was less important than the fact that this simple work allowed the child the pleasure of contributing to the family's welfare. When children believe they are contributors, they are less needy of repeated reminders that they are loved or of material signs that are supposed to persuade children of their value in their parents' eyes. I applaud the authors' suggestions to parents to treat their children as colleagues rather than as pedigreed racehorses that might one day win a large purse and celebrity.

Thus the authors affirm unequivocally that parents have a palpable influence on their children. The scientific evidence supports them. Children growing up in very different homes have different skills and motivations. Children from similar homes attending different day care centers are remarkably similar.

Finally, although the authors state correctly that parents should talk to, play with, and read to their young children, they also emphasize the importance of permitting the child to take pleasure from his or her own victories. The clever parent, like a good director of a play, guides the child but, at the last moment, provides enough freedom so that the child believes each achievement is his or her own personal accomplishment.

This book will answer many of the puzzling questions parents wonder about and, in so doing, will both deepen the affection most parents hold for their children and support the optimism that parents bring to this unique life assignment.

Jerome Kagan
Research Professor of Psychology
Harvard University

Acknowledgments

A s we send this book off to press, we would like to express our gratitude to all those who have inspired us, encouraged us, or helped us while writing this book. The list begins in order of appearance with our parents, Hilel and Ela Herschkowitz, and John and Clara Chapman, to whom we owe not only our genes but also the conviction that parents play an extremely important role in a child's life. We thank our children, Daniel Herschkowitz and Jessica Herschkowitz Gygax, for the experience we gained by watching them grow up.

David and Hillie Mahoney of the Dana Foundation were the first to encourage us to begin our project. Their enthusiasm and that of Barbara Gill and Barbara Best of the Dana Foundation and the Dana Alliance for Brain Initiatives cheered us on. We also thank Bill Safire for his steady support. Jane Nevins of the Dana Press has been our faithful guide and editor since the book's beginning. We thank, Mme Louise Demarest Thunin, writer, mother, and grandmother and Susan Aravanis and Elizabeth N. Lasley, both new mothers and writers themselves, for reading the manuscript and contributing valuable suggestions. Stephen Mautner, executive editor of the Joseph Henry Press, carefully co-edited the final manuscript. Randy Talley, production manager of the Dana Press, coordinated the art, and Kathryn Born, medical illustrator, prepared the illustrations for publication.

As representatives of the collective body of knowledge that has shaped this book, we have selected a few individuals who have been especially helpful by providing us with their recent publications, helped us with answers to specific questions, or shared with us their personal insights: Karl Zilles and Katrin Amunts, Brain Research Institute, Düsseldorf; Patricia Goldman-Rakic, Pasko Rakic, Yale University; Hennig Schneider, University Women's Hospital, Bern; Emilio Bossi, neonatologist and Dean of the Medical Faculty, University of Bern; Dennis Molfese, University of Louisville, Kentucky; Jean-Pierre Changeux, Institut Pasteur, Paris, France; William Damon, Stanford University; Joaquin M. Fuster, Brain Research Institute, University of California at Los Angeles; Carla Shatz, Department of Neurobiology, Harvard Medical School; Guy McKhann, The Zanvyl Krieger Mind/Brain Institute, Johns Hopkins University; Petra Hüppi, Department of Pediatrics, University of Geneva; Marcel Zentner, Department of Psychology, University of Geneva; Shintaro Okada, Department of Developmental Medicine, Osaka University, Japan; and Yoshikatsu Eto, Department of Pediatrics, Tokyo Jikei University School of Medicine, Japan.

Special thanks go to Jerome Kagan, Nancy Snidman, Eric Petersen, Steve Most, Sue Woodward, Marc McManis, and Doreen Arcus of the Harvard Infant Study and to Joseph Volpe of the Department of Pediatric Neurology, Harvard School of Medicine. We would also like to thank the staff of the Countway Medical Library; the University Hospital Library, Bern; and the Psychology Library of the University of Bern for their friendly and competent assistance.

A Good Start in Life

Introduction

Build *Your Baby's Brain.* As I was casually flipping through the vast array of compact discs in Tower Records on Massachusetts Avenue in Boston, a particular CD cover caught my eye. The image of the cheerful, alert infants, together with the claim that the "power of music" could build the brain was enough to pique my curiosity and send me off to the cash register. Acting on a similar impulse, I had already picked the September 13, 1999, issue of *Time Magazine* off the newsstand. The question boldly printed on the front cover was "The IQ Gene?" Did the article suggest that we could use genetic manipulation to make people smarter?

We all want to do the best we can for our children, and our experience tells us the importance of being able to handle vast amounts of new information in a rapidly changing world, of seeking creative solutions to the problems of our environment, of being able to deal with our emotions and get along with others.

Aware of the brain's central role in all of this, we are highly receptive to news from the field of brain research.

This is the Age of Information. However, it is also the Age of Confusion and Uncertainty. Waves of news are breaking against the shore with a splash of contradictory messages on what to do and what not to do. It is tempting for parents to grasp the nearest life raft and hang on for dear life.

Stepping back and listening to what science is telling you about how the human brain develops can help you strike a healthy balance between doing too much and doing too little for your child's cognitive, emotional, and social growth. You have probably wondered whether there are biologically determined periods in your child's development when you could miss critical "windows" of opportunity. You may have asked yourself whether there is such a thing as an IQ gene. Or you may have marveled at the way each baby is unique and wondered why this is so. Sociologists, psychologists, pediatricians, and neuroscientists are asking these questions too. We try to summarize their findings to give you the opportunity to add to your own knowledge and experience.

Our eagerness to absorb new discoveries should be tempered by a few general considerations. Not all study reports have immediate practical consequences. The fact that two events occur at the same time does not necessarily prove that one is the cause of the other. The brain is an immense network with numerous subsystems working together. For example, no one structure is alone responsible for memory or attention. While animal experiments provide us with valuable insights on basic neural processing and behavior, these findings are often still to be confirmed in human beings. The arrival on the scene of powerful, noninvasive techniques is now beginning to make possible direct studies linking structures to functions in humans. We are at the threshold of a very exciting time.

This book is a joint venture, based on our 34-year bilingual, American-European dialogue between neuroscience and education and between pediatrics and parenting. To simplify things, we decided to have me write the introduction, so that we don't have to call ourselves by name as if we were a third party looking on. Norbert's "voice" will then take over, but we are both actually writing this together.

Norbert and I discovered our common interest in child development on our first rendezvous—a tea under the trees at Stanford's Tresidder Union. Although our first page of notes for a book dates from 1967, many years passed before the book actually began to take shape. During this time, Norbert was head of the Department of Child Development at the Berne University Children's Hospital in Switzerland. This department was a multidisciplinary team including psychologists, psychiatrists, neurologists, and biochemists doing basic research on brain development. In the meantime, we raised two bilingual children, and I taught English to future primary school teachers. I marveled at how effortlessly young children learn to understand and use speech and shift smoothly from one language to another. I was often puzzled as to why students had difficulties with even basic features of a foreign language. Living in Europe, with annual stays in the United States to visit family and work with American colleagues, has made us aware of differences in attitudes about parenting and education. But more important than the differences are the similarities in all parents' hopes and fears about their children's future.

As a mother and a teacher, I followed Norbert's work in neuroscience and pediatrics with special interest. While we worked together on articles for scientific journals, attended congresses, and prepared lectures for the public, I became aware of what was going on in brain research and of its relevance to child development. It has not been easy for me, a nonscientist, to understand basic research. Nor has it been easy for the scientist to find the right language to communicate neuroscience to a layperson like me, more interested in novels than in neurons.

Two events combined to propel our ideas for a book into the foreground. About seven years ago, Norbert began to work with Jerome Kagan, of the Department of Psychology at Harvard, and Karl Zilles, of the Brain Research Institute, Düsseldorf, on a study of the development of brain and behavior in infancy. This work opened new dimensions for us in the fields of child psychology and brain biology. We are extremely grateful for this experience and for the warm personal relationships that it has led to.

The second momentous event was the day we met David and Hillie Mahoney, of the Dana Foundation, New York. Their enthusiasm for promoting a dialogue between scientists and the public was contagious, and we were convinced that it was time to make our book a reality.

Nature has equipped human beings with an instinct to protect and nurture infants but has not provided us with universal rules for parenting. Parents must, therefore, adapt practices handed down to them by members of their surrounding community, taking into consideration the individual child and the family's needs and expectations, as well as those of the wider society. Your common sense, together with knowledge of how your child's nervous system develops and interacts with her environment, is a valuable guide.

The brain plays an active role in everything we think, in everything we feel, in everything we do. It is the organ that interprets the multitude of signals from our sensory organs and forms associations among them, a basis for memory, learning, and actions. At the same time, it is the seat of our emotions. Hippocrates, in ancient Greece, called attention to the fact that "the source of our pleasure, merriment, laughter, and amusement, as of our grief, pain, anxiety, and tears, is none other than the brain." Hopes and desires arise in the brain, as do the strategies we evolve to reach our goals.

Because our control centers for vital functions such as blood pressure, respiration, reactions to infections, and the release of hormones are also located in the brain, it is involved in our general health and how we cope with challenging situations. New research on the sensitivity of emotional and stress systems illuminates the importance of the individual personality in the brain–body connection.

Reports of the intense development taking place in the child's brain have led to the assumption that a child's first years are a time of critical windows of opportunity that are decisive for the child's entire future life. However, as John T. Bruer points out in his book *The Myth of the First Three Years*, it is helpful in regard to critical periods to make a distinction between development that takes place automatically in the presence of stimulation naturally available to all human beings and development that depends on cultural surroundings and specific individual experience.

Nature builds the human brain according to a universal genetic blueprint and general timetable—with a wide range of individual variation—to meet the conditions of the world a baby finds outside the uterus. We don't have to teach babies to see or hear. Babies figure out for themselves how to sit, stand, or walk. And they pick up the language spoken around them. The stimulation they find around them is sufficient for their systems to develop, as long as their sensory organs and nervous system are intact. Here critical periods may be missed. We know, for example, that the baby's later ability to see objects in depth depends on normal visual experience in his first two years. Visual deficits must therefore be corrected early.

Other capabilities depend on specific experience, learning either by example or by direct instruction. Just how this is done depends very much on the needs of the child's cultural surroundings. All cultures evolve norms for handling emotions and getting along with other members of the society. In addition, each society has its particular needs. Before the invention of the printing press, a valuable skill was being able to remember and retell long passages of text. Later, reading and writing took that skill's place. In today's world, the ability to use computers and deal with the immense amount of information they provide has become of utmost importance.

With respect to these culturally dependent skills, modern brain science brings good news. Their acquisition does not depend on biologically determined "critical" periods—in the sense that if they are not learned by a certain age, the brain will never be able to make the necessary connections later. Nevertheless, we can speak of periods that are especially fertile for learning; that is, phases when it is indeed easier to acquire new skills. It takes less effort, for example, to learn a second language before the age of 12 years than it does in adulthood.

Your child's first years influence but do not determine the person he will be later in life. As he grows up, his surroundings will expand to include more influences outside the home and family. At the same time, he will make his own selection, based on his own unique combination of genetic traits and personal experience, from the opportunities offered by the environment. He will select other role models and pursue new idols and ideals. His personality and his brain will both develop over his entire lifetime.

However, each stage of development has an impact on a child's transition to the following stage. The early years of your child's life are special because a vast amount of unconscious learning takes place, as attitudes and habits are formed, mainly with you as role model. This has an effect on his or her start in life outside the family. When a child's initial experiences on the playground and in school are positive, it serves as an incentive for further learning and the development of social competence.

There is a lot you can do. You can stimulate your children's curiosity and imagination; you can help them learn to overcome frustrations and evolve strategies for solving problems and reaching goals, to experience problems as "challenges" that can be met, not merely as "stress" that they cannot cope with. You can encourage your children to develop language abilities that will help them to learn more about their world and communicate with others. You can help them understand that their actions—or non-actions—have consequences. You can strengthen their sense of accomplishment and build up their sense of responsibility. You can foster empathy for the feelings and needs of others and can widen your children's horizons from their own concerns to those of their family and of their community. Adults can raise a child's self-esteem by making him or her feel a part of "us" and not just an isolated "me."

Our book begins with a short account of your baby's life in the womb and the impressive development of brain and behavior that goes on during pregnancy to prepare the baby for life in the world. In the three parts that follow, we discuss chronologically your child's emerging capabilities together with the current state of knowledge of human brain development.

In addition to providing information on human brain development—those processes that take place in all children, generally in a similar time frame—we tell of new scientific evidence that helps to explain a child's individuality. We include practical examples of how you can better understand your child's personality and follow her individual path of development. However, the material in this book does not replace a pediatric assessment of your child's physical, cognitive, emotional, and social development.

Because our heads are full of the real children that Norbert and his colleagues have treated and studied over the years, we found it hard to think of them in an anonymous, one-size-fits-all way. So we settled on the pleasure of creating fictional children like all the ones who've crawled, toddled, and marched through our lives. The story of the young parents, Allen and Deborah, their children, Emily and Andrew, and Emily's small circle of playmates illustrates the variation in developmental timetables and the broad spectrum of individual personal characteristics observed in children.

We use the word *parents* in the text to include all persons involved in a child's care. It is particularly interesting that human beings are probably the only species in which grandparents play a role in bringing up children. Because babies come in both sexes, we have used *he* and *she* in alternating chapters to describe behaviors that are typical for all babies.

Each chapter concludes with a section called "To Think About." Here we bring up questions raised by patients or audiences at public lectures, and we address topics of current interest for daily life, for example: Can prenatal stimulation enhance brain development? How important is early schooling? Why are children in one family so different? The comments in this section are related to material discussed in the chapters and are based on current knowledge in this field.

To help you find your way through the intricate network of the brain, we have provided a set of "brain maps" in the center of the book showing the names of structures or regions in the adult brain. You can refer to it and the accompanying glossary whenever an unfamiliar term appears in the text.

It is our sincere hope that knowledge about the way your child develops and interacts with his or her world will help you establish the warm, supportive, and challenging environment that stimulates the child's growth as an individual and as a member of society and, at the same time, contributes to your enjoyment of your mutual voyage of discovery.

Elinore Chapman Herschkowitz

Part One
Getting Ready

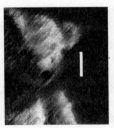

Life in the Womb

The test is positive! Until now, you have been listening politely as your "nouveau-parent" friends regaled you with their triumphs and tribulations with their enchanting offspring. But now you suddenly begin to picture yourselves in their position. What will your own baby be like? Will it be a rough-and-tumble boy who will play the trumpet, join you in a game of chess, accompany you on long hikes in the woods? Will it be a girl with organizational talent who takes over the family business? You are sure of one thing: you are determined to do everything in your power to give your new family member the best possible start in life.

During the very first weeks after conception, before you even know for sure that your baby is more than just a dream, she has already accomplished some big steps on her way to becoming a real individual. At around 20 days, the baby's heart muscles start beating. A few days later, the first signs of the baby's future arms and legs appear. The baby's brain gets off to an even earlier start, and it needs to. It has a big job ahead, namely, to prepare the baby for life outside the uterus in the short span of only nine months.

The Brain Takes Shape as It Goes to Work

By the embryo's second to third week of existence, when it is only about one-eighth of an inch long, the brain begins to form. (Biologists and medical personnel use the term *embryo* to describe the new human being up to the end of the third month and *fetus* for the remaining time before birth. Parents, understandably, prefer to use the word *baby*.) As in an artist's preparatory sketch, the faint outlines of the future structures of the brain emerge before the details are filled in. The basic design and the timing of construction follow the instructions contained in the baby's genes. While some of the structures will take up their functions before birth, others will gradually join up later.

At the end of the first month a crevice appears, separating the future left and right hemispheres. These visible halves of a single sphere are the source of fascinating and often fanciful "left brain–right brain" theories about our personalities and dispositions. While the two halves will gradually become more specialized for performing particular functions, they will always continue to act together and complement each other.

Until around the seventh week, the baby's brain and body are built up according to a sexually unspecific pattern. However, if the baby's genes contain instructions for a male, a special factor will now trigger the formation of male sex organs. The sex organs begin to produce the hormone testosterone, which will exert its influence on the developing brain structures, making the male brain different from the female brain and thereby affecting the timing of brain development.

For all the rapid construction taking place in the brain, a tremendous number of nerve cells, the basic building blocks of the nervous system, have to be formed. Between the baby's second and seventh month in the uterus, more than 100 billion nerve cells, called *neurons*, are formed, roughly half the number of stars in the Milky Way. This means that at times more than 250,000 neurons are formed per minute. We used to think that all the nerve cells we would ever have were generated before birth. However, recent research suggests that some limited nerve cell formation takes place even in adult life.

From the moment they are formed, the immature nerve cells migrate to their genetically predetermined locations in the baby's brain. To give you an idea of the magnitude of the distances involved, the neuroscientist Pasko Rakic has said that if nerve cells were the size of people, it would be as if the whole population of the United States were to migrate from one coast to the other. The first cells to form stake their claims on the closest sites, leaving the latecomers to blaze a trail through them to find their final destinations. It is a miracle that so few neurons go astray on their long march through the crowds. However, when many cells do get lost—for example, as a result of infections—it may be a factor in developmental disorders such as cerebral palsy, epilepsy, mental retardation, and autism.

Along the way, the neurons start to take on the specific form that will enable their specialized functions later. They grow branches called *dendrites*, which receive signals, and an extension, or *axon*, which transmits the signals further. Upon arrival at their destinations, like students at a new school, the neurons begin to actively seek contact with other neurons and form connections, called *synapses*. The famous Spanish neuroscientist Ramon y Cajal romantically referred to synapses as "protoplasmic kisses." At the synapse, the axon of one neuron meets the dendrite of the next neuron without quite touching it. Chemical messenger substances called *neurotransmitters* carry the cell's message across the gap.

Around the seventh week of the baby's prenatal life, neurotransmitters are already detectable. These may play an important role during the early phases of brain development by stimulating the growth of the brain structures and only later function as actual "messengers." A few weeks later, synapse formation begins in earnest. However, most synapses will be formed after birth.

Forming synapses is a matter of life and death for the cell. Neurons that fail to make a connection wither and disappear in a process known as *apoptosis*. While apoptosis may sound tragic for the single neuron, it is essential for brain development and allows more resources for important connections.

First Movements

Imagine yourself lying quietly in bed, thinking over the fact that the first half of your pregnancy is about over. Suddenly you feel a vague, bumping sensation in your abdomen, like something small pressing against the wall of your uterus. Yes, it moved. There it is again! You call the baby's father, and if he's quick enough he can reach over and feel the movement with his hand.

Until recently we could only imagine what babies are doing within the warm, dark confines of the uterus. Now modern ultrasound techniques provide new insights into the baby's activities, even if they don't give us clear pictures like closed-circuit television cameras. Ultrasound is used in routine examinations and is not harmful for the fetus.

We now know that a baby begins to move long before her mother can feel her motions. As early as about the end of the second month after fertilization the muscles of the tiny embryo begin to twitch. The muscles react asymmetrically. Those on the embryo's right side are more active than those on the left.

At around three months the baby begins to perform motions that look more like "practice" for postnatal life. Instead of moving and twisting her whole body at once, the fetus begins to move her limbs singly—one arm or one leg at a time, for example. Again, most fetuses move their right arm more than their left.

Single-limb movements are possible because nerves from the baby's spinal cord have now grown long enough to reach and make contact with the muscles. The nerves in the baby's spinal cord are now able to send signals to the muscles, causing them to relax or contract.

The baby is not only exercising her arms and legs but is also practicing many movements that are essential for eating, drinking, and breathing. She begins to open and close her jaw, move her tongue, and make sucking and swallowing movements. Movements can be detected that resemble a prenatal yawn and hiccup. A few weeks later, a fetus can even suck on her thumb, and we're not surprised to hear that it is most often the right thumb. These movements are a sign that the baby's brainstem is now an active member of the motor team.

With respect to mouth movements, Peter Hepper and his colleagues at Queen's University in Belfast observed interesting gender differences in fetuses at the tender age of four to five months. Using ultrasound techniques, they saw that at four months male and female fetuses made about the same number of mouth movements. But a few weeks later the females were way ahead of the males. No long-term conclusions about future verbal abilities or conversational habits should be made on the basis of this finding. However, it is an indication that females are ahead of males in brain development by this time.

Around the time when a pregnant woman first becomes aware of her baby's activity, at about five months, the fetus's movements become more coordinated and smoother than before. This is because new structures now join up with the motor team, putting movements of the muscles under increasingly higher levels of control. At around four months after fertilization, the baby's cortex is just beginning to exert its influence. This is the time when neurons from the baby's motor cortex, the part of the cortex that controls muscle movements, make connections with the neurons in the spinal cord.

Feeling My Way Around: The Sense of Touch

As a baby grows in the uterus, her senses are already busily receiving information from both outside and inside her body. The stimulation inside the uterus is just what the developing systems need. A baby's sense of touch develops early and gets a lot of practice before birth. As a fetus moves, she bumps against the uterine wall, and sometimes her tiny hands brush up against her face. Between two and five months after fertilization, touch receptors develop in the baby's skin. The first to appear are those around the mouth, and they will be more numerous in this region, which is why babies explore so many things with their mouths. Although the touch receptors register a sensation automatically, they are not yet linked to higher centers.

Between five-and-a-half and seven months after fertilization, contacts are established between the touch receptors and cells in the baby's somatosensory cortex, the part of the baby's cortex that is specialized for touch sensations. *Somato* is derived from *soma*, the Latin word for "body." After an initial processing in the so-called primary area of the somatosensory cortex, the information from the touch receptors is passed to higher cortical areas, which put the information together. Now the baby becomes able to register the sensation of touch.

Together with maturational changes in the brain, the fetus's experience with touch stimulates the growth and specialization of the neurons in her somatosensory cortex, making them ever more efficient at processing touch sensations. Around this time, it is likely that a fetus could also feel pain.

Can You Hear Me in There?

The evangelist Luke, who was also a physician, showed a good knowledge of fetal behavior when he wrote: "And when Elizabeth heard the greeting of Mary, the babe leapt in her womb" (Luke 1:41). This was around the sixth month of Elizabeth's pregnancy. Contrary to the opinion of Luke, medical authorities still claimed until the early part of the twentieth century that unborn infants weren't able to hear. In 1925, Dr. Albrecht Peiper, a doctor in the Berlin University Children's Hospital, made an interesting observation. He played a few notes on a toy trumpet to some newborn babies—the youngest was just 25 minutes old—and noticed that their body movements changed.

Intrigued, Peiper decided to see if babies could hear while they were still in the uterus. Women in the last weeks of pregnancy were asked to lie quietly so they could feel the movements of their babies. Peiper wanted to exclude the possibility that the mother's startle or change in breathing had an effect on the infant's behavior, so he gave a warning count before he blew a loud blast on a car horn. Some of the babies reacted with a thump against the wall of the uterus; others wriggled around for a while. After he repeated the sound several times, the babies reacted less. Peiper made a note that the babies behaved consistently on later repetitions of the experiment, but he didn't mention that this might reflect the baby's individual style.

Ultrasound techniques and heart-rate measurements confirm that fetuses begin to respond to sound at around five months. They startle at a sudden noise, blink, and stop moving around, and their heart rate decreases momentarily. Around this time, the baby's cochlea, or inner ear, begins to function. This is the time when nerve fibers grow out from the thalamus to make their connections in the baby's auditory cortex.

How much does a fetus actually hear? She is surrounded by the sound of the swishing noises—doctors call them *borborygmi* (bor-bor-IG-me, rhymes with pygmy)—of her mother's intestines. The general sound level in the uterus can reach about 70 decibels (dB), similar to the level we perceive when we run a vacuum cleaner. Her mother's heartbeat can be felt as vibrations. Her mother's voice rises about 24 dB above the background sounds and is also transported directly as vibrations, so by her last month in the uterus, the baby can definitely hear it. Studies have shown that a whole month before birth, fetuses are able to distinguish among music, heartbeats, and speech sounds.

During "our" sixth month of pregnancy, we were walking in San Francisco at Chinese New Year time. All around us, we heard the nervous popping of firecrackers on the sidewalk. Elinore felt that each volley of firecrackers was met with a vigorous drumming from inside her uterus. Our son didn't grow up to play a percussion instrument, but we often joked about his fascination with fireworks.

The View from Inside

The view really isn't much, perhaps at most a faint orange glow during the last weeks before birth. But even in the dark, your baby's visual system is under intense construction to prepare her for her life in the world of light. Already at around one month after fertilization, when the first traces of her brain as a whole come into view, tiny bulges that will become the baby's eyes appear.

Carla J. Shatz, now chair of the Department of Neurobiology at Harvard Medical School, showed in animal experiments that the basic wiring of the visual system begins to take place before any stimulation from the outside world reaches the baby's eyes. In the absence of light, special nerve cells in the retina of the eye called *ganglion cells* begin, probably under genetic influences, to fire off short bursts of electrical impulses. The impulses are transmitted from the retina along the optic nerve to the brain. The spontaneous electrical activity of these retina cells seems to be crucial for setting up the correct wiring. If it does not take place, vision will not develop normally.

The impressive groundwork takes place all by itself without any extra outside stimulation. However, adverse environmental conditions may prevent necessary developmental steps from taking place. The spontaneous firing of the nerve cells in the visual system is vulnerable to disruptions. Drugs that interfere with the transmission of electrical activity across the synapse (e.g., nicotine, benzodiazepines, or narcotics) could disturb the pattern of the fine connections and lead to later visual deficits.

By the time your baby is born, her visual system is basically set up. But it will need the stimulation of the outside world to complete the job—and there is plenty out there waiting.

What's for Dinner?

You might have heard that newborn babies recognize their own mother's milk and sometimes even turn up their noses at it if she has tried some unusually spicy food. This means that the systems for taste and smell are well developed by the time the baby is born.

Taste and smell are similar in that they are "chemical senses." That is, they involve direct contact with molecules of the chemicals that make up the substance the fetus perceives through her mouth or nose. Special receptor cells convert the chemical into an electrical signal. After passing through a series of relay stations, the signal reaches the baby's cortex, where it can be registered.

Sometime between two and three months after fertilization, taste buds appear on the edge of the baby's tongue, on the roof of the mouth, and in the upper throat area. Some taste buds and their connections to the brain are already functional by the third trimester of pregnancy. But they become more numerous and continue to develop throughout the rest of pregnancy and even for a few months after birth. Evidence that babies' taste systems develop early is the fact that babies born prematurely at 24 weeks already show a basic sense of taste.

At around four to six months after fertilization, the plugs blocking the baby's nostrils disappear, making it possible for chemical substances in the amniotic fluid to come in contact with the receptor cells in the baby's nose. During the third trimester the olfactory receptors for the sense of smell are developed. Although your baby can't yet share your dining pleasure, some signals—strong flavors such as garlic—may be getting through.

Learning in the Womb?

We know that while babies are growing in the uterus they are stretching and turning and that their senses are diligently processing incoming information. But are they able to form memories? And if so, do these memories affect the baby's behavior; that is, do they represent an early form of learning? Dr. Albrecht Peiper, of the car horn, did not agree with the many scientists of his time who believed that the capability to form memories begins only after birth. He had observed that fetuses reacted intensely when they first heard the car horn. But after they heard the horn several times, they reacted less and less and eventually stopped moving around. He proposed that the sound left a trace that inhibited further reactions to the sound and wondered if this *Merkfähigkeit*, or ability to remember a stimulation, was an early form of memory.

Later studies have confirmed and extended Peiper's observations. The "memory trace" that Peiper referred to is now called *habituation*, which is considered one of the simplest, yet essential, learning processes. Habituation, which will play an important role all through life, is the ability to react less strongly to a repeated stimulus. After their initial startle, the babies got used to the sound of the car horn. Ultrasound techniques have demonstrated that fetuses as young as 23 weeks after fertilization are able to habituate. Habituation appears first in females, a further indication that the female nervous system is on an earlier schedule than the male.

To find out how long fetuses can keep a memory, Peter Hepper played a particular tune to mothers during the ninth month of pregnancy. He played it loudly and often enough that the fetuses had a good chance to become familiar with it. One week after birth, the babies behaved differently when they heard the familiar tune from when they heard one that was new to them. But two weeks later, they didn't distinguish between the familiar and unfamiliar tunes anymore. This shows that the fetuses could form a memory, but it was only of short duration.

We know that in the mature nervous system a structure called the *hippocampus* is important for memory formation. It is, therefore, of interest that this structure undergoes intense development at around five to six months after fertilization. However, to store a memory for a longer time—over months, years—the brain has to transfer it to the baby's cerebral cortex. This transfer does not appear to be taking place at two weeks after birth.

The current knowledge about how memories are formed and learning takes place suggests that it is unlikely that unborn babies form lasting memories. However, basic processes involved in

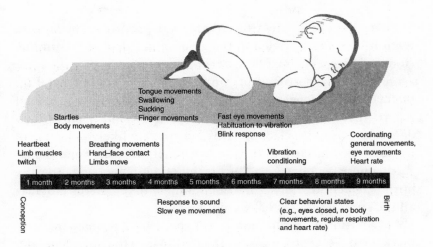

The timetable for development of fetal activity is very similar for all babies, unlike the development of behaviors after birth, which varies widely.

memory and learning are under way during the baby's time in the uterus, forming the foundation for a running start at birth.

Prenatal Stimulation?

Even before people had any idea of what babies are doing in the uterus, they considered trying to influence the child's development. Talmudic writings from the second to sixth century contain references to prenatal stimulation programs. It would be interesting to know their "curriculum."

When Frank Lloyd Wright's mother was pregnant, she hung up pictures of English cathedrals and spent a lot of time looking at them in the hope that it would have an effect on her child. As we know, Frank did go on to become an architect and, in fact, announced early in his career that he intended to become the greatest architect who ever lived. However, the reasons for his success can be found in the combination of Frank's own talents, his personality, and his mother's encouragement after he was born. If a mother's thoughts alone were enough, how different parenting would be!

In recent years, scientific reports on the ability of the fetus to learn have sometimes led to the assumption that early stimulation might speed up brain development and lead to greater achievement in postnatal life. In his 1992 review of prenatal influences, Peter Hepper notes that there have been some anecdotal reports that prenatal stimulation has an effect, but scientific evidence for this idea is lacking. Any benefits are likely to be due to the increased interest of the mother in her pregnancy and the resulting positive effects on her lifestyle both before and after the birth of her child. The natural environment of the uterus provides all the stimulation the baby's brain requires.

If additional stimulation could speed up development, premature babies, exposed earlier than normal to the outside world, would be ahead of babies who spent the full term in the womb. But our own studies show that, in spite of their earlier exposure to outside stimulation, babies born after only 32 weeks do not have a head start over full-term babies. Dr. Petra Hüppi, of Boston and Geneva, Switzerland, used behavioral scales and magnetic resonance imaging (MRI) techniques to monitor and compare the development of premature and term-born infants. She compared the preterm babies at their expected due date at 40 weeks to term babies born at 40 weeks. In spite of their eight-week head start, the premature babies' development was delayed. The eight weeks that the term babies had spent in the uterus were apparently beneficial.

Although the development of brain structures is not speeded up by extra stimulation before birth, negative influences can impair a baby's brain development. Alcohol, nicotine, drugs, and malnutrition are known risk factors, as are X rays and some infections.

Does a Mother's Stress Affect the Baby?

If you are racing around from one appointment to another, struggling to balance school schedules, deadlines, and menu plans, to say nothing of trying to find a quiet moment to collect your thoughts, you may worry about what effect your daily hassle could have on your baby's brain. As early as 480 B.C., Empedocles suggested that the development of the embryo could be guided and interfered with by the mental state of the mother. And 1000 years ago in China, "prenatal clinics" were established to keep mothers tranquil, which was thought necessary to maintain the psychological health of the fetus.

In the early 1970s, scientists began to approach this question in a systematic way. Researchers in Canada studied the relationship between a mother's stressful situations during pregnancy and the baby's postnatal development. The investigators found that mothers who suffered constant high, personal tensions during pregnancy, mainly marital discord, had children who had an increased risk for eczema and tended to reach motor milestones later than infants of mothers whose pregnancies had been more relaxed. The babies also tended to be more fretful and restless and to have difficulty quieting down. The authors suggested that changes in the endocrine system of the mother resulting from stress could affect her unborn child.

Everyday life is full of short-term events that, depending on your personality, may be perceived as "stressful": you are frightened by a sonic boom, your two-year-old races out into the street, your boss criticizes the report you worked on all weekend. Your nervous system reacts to these sudden events by causing a surge of adrenaline into your bloodstream. This surge may lead to a restriction of blood flow to the uterus, similar to the effect of a mother's smoking. The fetus detects the change and "sympathizes" with your upset condition: her nervous system also produces more adrenaline, causing a temporary change in heart rate and body movements. Another way in which maternal mood can affect the fetus is through the hormone cortisol, which mobilizes the body in times of stress. If the mother is highly anxious, her body produces more cortisol. Some of this passes directly to the fetus, and some acts over the placenta to stimulate the baby's own endocrine system. Sporadic bouts of stress or low-level "worry" have no long-term effect on the baby's development.

Prolonged and severe stress during pregnancy, however, may have consequences. Reduced blood flow to the placenta may restrict the baby's growth and lead to the condition known as "small-for-date." Recent studies have linked high levels of mothers' reported anxiety, particularly during the latter part of pregnancy, to a cluster of behavioral problems that appear in early childhood. These include hyperactivity/inattention, particularly in boys, and emotional problems in both boys and girls.

It is important to realize, however, that the infant's brain is constantly shaped by her experiences after birth. This gives parents the opportunity to counteract the effects of the prenatal stress by adjusting their childcare techniques to meet the needs of the child. This could mean, for example, providing a calm, "predictable" environment and avoiding excess stimulation.

Signs of Individuality

My neonatologist friends have frequently told me that babies react very individually to routine ultrasound procedures. Some get excited and kick back at the ultrasound head, while others remain quiet. These differences might have something to do with a baby's temperament. Knowing more about this pattern would help us to learn about a baby's early characteristics that are not influenced by her experiences after birth.

A 1999 study confirmed what mothers have experienced all along: that babies show individual patterns of activity while still in the uterus. At the middle of the eighth month, investigators measured the rate of the fetus's general spontaneous movements on three different occasions using ultrasound techniques. Each child showed a clearly individual pattern of body movements. The researchers then compared these data to the baby's activity during the second and fourth week after birth. The babies who moved more often or kicked around with greater energy before birth were the ones who were also more active during their first months outside the uterus.

Ready to Go

The 40 weeks of pregnancy seem long enough to you, but it is an amazingly short time for all the intense building activity that has to go on in your baby's nervous system. By birth, the production and migration of the nerve cells are practically over. Brain structures are in place, and the major connections are functioning. While the brainstem, the structure responsible for vital functions, is practically fully developed, other structures undergo major construction after birth. The main bridge, or *corpus callosum,* is beginning to link the brain's two hemispheres. Neurotransmitters are being synthesized. Electrical activity is going on.

The preferential treatment that a baby's brain enjoys at this early stage is illustrated by the fact that her total birth weight is only 5 percent of what she will weigh as an adult, while the weight of her brain is already 30 percent of adult brain weight. Additional evidence for the brain's importance at this time is the amount of energy it needs. While the adult brain uses only 20 percent of the body's total energy supply, the newborn's brain consumes almost all of it.

Networks involved in vital functions, such as breathing and circulation, are ready to go into action immediately at birth. Your baby's sensory systems are ready to take in the wealth of stimulation awaiting her in the outside world. She has been moving her trunk and limbs by contracting and relaxing her muscles in practice for her new life outside the womb. Her sense of touch is ready to respond to your caress.

The ability to transport fragile memories formed in the uterus over the bridge into postnatal life will help your baby feel at home in her new surroundings. The melody of your voice and the scent of your body help the baby find nourishment and a haven of comfort and security.

To Think About

Will extra stimulation help my baby's brain development?

We now know that life in the womb offers a great deal more stimulation than we might think at first. This stimulation is indeed important for the development of the embryo and fetus. But there is no proof that more is better. If you find listening to Mozart enjoyable and relaxing, if reading Dr. Seuss aloud is practice for reading to your child later, or if prenatal programs make you more interested in your child's development, this kind of "indirect stimulation" is most likely beneficial.

Should I eat for two?

Even if you don't need to eat twice as much during pregnancy just because you are eating for two, you do need to consider the additional requirements for the growing baby. Consult your doctor about a supplement of folic acid. Be sure to eat a balanced diet with all the foods that are good for you, too: fresh vegetables, milk or soy products, hearty whole grains, luscious fruits.

Which substances have been proven harmful to a baby's brain development?

Some substances are known to have negative effects on a baby's brain development. It's certainly best to avoid alcohol, nicotine, and illegal drugs entirely. In high doses, X rays and radiation can lead to stunted brain growth. Becoming overheated, especially during the first trimester, is to be avoided. Treat fever with a drug approved by your doctor, and stay away from hot tubs and saunas. Immunization against chicken pox and rubella should be undertaken before you become pregnant. Avoid organic solvents such as toluene or benzene, or use them only in well-ventilated rooms.

Are modern electronic devices dangerous?

To balance the list of dangers, it's a relief to know that many modern devices such as video displays and microwave ovens have not been proven to be harmful. Neither ultrasound nor MRI procedures have been shown to be a risk for the infant. However, MRI should be avoided in first trimester until more is known.

When is caution necessary?

A pregnant woman should consult her doctor or other health care provider before taking any drugs or medication of any kind, including herbal remedies. Although they may look "natural," many of these contain unspecified amounts of substances that could be harmful. If you can't do without that morning cup of coffee, limit your coffee intake to no more than two cups a day.

Since influenza presents a possible risk to the fetus, take care during the flu season. A simple but effective means of reducing the chance of contact with the flu virus is to wash your hands frequently and avoid touching your mouth or nose.

In 1997 the American Academy of Pediatrics warned that excessive noise could have harmful effects on the development of the unborn child. Studies have shown a correlation between high noise levels and high-frequency hearing loss in newborns, increased risk of preterm delivery, and decreased birth weight. However, more studies are necessary to determine whether the noise itself was the cause or whether the problem is also related to unfavorable socioeconomic conditions.

What is meant by "excessive noise"? A general recommendation is not to be exposed to much above 80 dB for any length of time. To give you an idea, this level is about the noise of heavy traffic. A power mower is about 100 dB, and the noise really becomes "excessive" with a boom box in a closed car—about

120 dB! Jet planes overhead can bring the noise level up to 140 dB. In addition to the effect of a general high noise level on the baby's hearing system, unpredictable loud noises, such as the roar of planes taking off or landing at a nearby airport, can be stressful for the baby's mother.

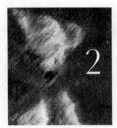

2

Birth

During the baby's passage through the birth canal, the umbilical cord is squeezed, reducing the supply of blood from his mother. The oxygen level in the baby's blood goes down, and the carbon dioxide level rises. When the baby is born, the drop in oxygen makes him take a deep breath, causing his lungs to unfold and fill to full capacity. When the freshly expanded lungs suddenly release their contents, the air rushes through the narrow channel of the larynx and produces the baby's dramatic announcement of his presence, his first cry: "Here I am!"

The Brainstem's Big Moment

Babies arrive with a bright-eyed, alert expression, as if they are eager to become acquainted with their new world. This is also the work of the brainstem. When neurons in the brainstem are excited—by loud sounds or strong patterns of light, for example—the neurons produce a sudden surge of norepinephrine, a neurotransmitter that makes the infant more alert and ready for action. Thanks to the baby's breathing, his brain gets five times more oxygen than it had in the womb.

A popular word today is multimedia, but the brainstem has been in that business for a long time. The brainstem receives incoming sensations through separate channels, or "modes." Information from the eyes is in the "visual mode," what a baby hears is in the "auditory mode," and what he feels by touching his skin is in the "somatosensory mode." His brainstem integrates the input from the different channels, a multimodal process.

The brainstem is also "interactive." When a baby hears a sudden loud noise, he startles. The muscles of his whole body jerk in unison. When he sees a face nodding above him, he moves his head and eyes to follow it. His brainstem receives the signals and automatically activates the appropriate muscles.

Between midgestation and birth, an insulating sheath called *myelin* begins to form around the axons of the neurons in the brainstem, greatly increasing the speed and efficiency of the signals. The fact that myelination takes place so early in the brainstem is evidence of this structure's great importance. The intense activity going on in the brainstem around birth means that it needs extra energy right now. When brain cells are particularly active, they take up greater amounts of glucose, their main fuel.

This is the brainstem's big moment. Around the time of birth and during the following three or four months, the brainstem performs its role as central dispatcher with little interference. However, the brainstem is linked to other structures, such as the hippocampus and cortex. These structures are developing rapidly and the connections between them are growing stronger. During the coming months, the cerebral cortex gradually assumes its role in modulating the activity of the brainstem.

First Things First

Mixed with our joy in welcoming the new arrival is concern for how well the baby has navigated his arduous passage and how ready he is to cope with the conditions of life outside the uterus. Although birth is a natural process and takes place in most cases without complications, some infants have problems in adapting to their new life. Thanks to advances in perinatal medicine, it is now possible to monitor the infant's vital functions, such as heart activity, even before birth and support the baby's adaptation to his new surroundings.

In 1952, Virginia Apgar, an anesthesiologist at Columbia University, presented a system that initially assesses how well the baby is adapting to life outside the uterus. Through her work in the emergency room, she had become convinced that detecting problems within minutes of birth could prevent serious damage to the baby's brain. The baby's heart rate, respiration, muscle tone, reflexes, and skin color are noted at 1, 5, and 10 minutes after birth, and the results are converted to a numerical score. A score of below seven means that the baby needs medical help, and a score of below four means that intensive care is necessary. It is important to realize that the newborn assessment procedures are not intended to predict the future development of the infant. Their purpose is to determine which infants are at risk and require immediate medical attention. Apgar's scoring system is now used in delivery rooms all over the world, and a recent study showed that it is as valuable today for predicting neonatal survival as it was at its introduction half a century ago.

While still in the delivery room, the attending nurse or doctor weighs the baby and measures his length and head circumference. Body measurement gives a general picture of the development of the newborn. Are the baby's body weight and length appropriate for the time he spent in the uterus? When the baby's weight is significantly lower than expected at term, it is a sign of slow prenatal growth, which can occur for a number of reasons, such as malnutrition of the mother, insufficiency of the placenta, infections, or exposure to drugs. When the baby is too light and too small but his head circumference is appropriate for his gestational age, it indicates that even under adverse conditions leading to intrauterine growth retardation, the brain was still given preference. But when the head circumference is also too small for the gestational age, it tells us that brain growth may have been affected as well.

I became a father for the first time when I was teaching students at medical school how to examine a newborn baby. As soon as our son was born, I raced him out of the delivery room to a nearby examination table and went through my whole neurological checklist, about 42 items. I then proudly went up to my wife's room (those were the days before "rooming-in," which now allows mother and baby to be together right from birth) and gleefully exclaimed that I had checked the baby and he was fine. My wife looked at me and asked, "What color are his eyes?" I had to go back to the nursery to look.

The Newborn's Senses

Babies greet the world with open eyes, with open ears, and with open arms. For many weeks in the uterus, your baby's eyes, ears and sense of touch have been practicing for this moment, and now these senses are ready to immediately begin their task of assimilating and reacting to the vast amount of new impressions awaiting him upon arrival. Babies put all their senses to work to get their bearings on their new environment.

The newborn's world may be all new, but it is not the chaos it would be if an infant were not able to extract contours and shapes from the multitude of visual sensations, or if he made no distinction between speech and noise. The world would be a mass of meaningless light impulses and sound waves. Instead, his brain comes with highly specialized systems to process and store incoming information. Exciting new research on brain development combined with more precise techniques for observing babies' behavior is expanding our knowledge of how a baby responds to his new world.

Looking around

Once your baby has taken his first breath or two, his eyes blink at all the bright light in the delivery room and he stares astounded

A newborn's view of a face compared to the view of a six-month-old.

by all the unfamiliar sights. Now his sense of vision suddenly becomes important for helping him become acquainted with his world. While his sense of hearing has been exposed to a great variety of sounds in the uterus and his sense of touch has been stimulated by contacts with his own body or the uterine wall, his visual system has been pretty much left in the dark until now.

Newborn babies seem eager to find out what's going on. Marshall Haith and his colleagues observed infants lying in a dark room. Even in complete darkness, the infants' eyes moved around as if they were looking for something to attract their attention. Since no light was entering the babies' eyes, the investigators concluded that the eye movements were "endogenous," that is, the result of direct activity of the brain rather than of outside stimulation.

A newborn baby sees faces as blurred figures surrounded by areas of light, and the baby's eyes can focus only on objects that are within about 8 to 30 inches. This is about the distance between a mother's face and that of her baby when she is holding him during feeding.

Newborns focus on strong lines and distinct contours. Studies have shown that they can tell the difference between the outline shapes of a triangle, square, circle, and cross. In human faces, the eyes and hairline are prominent features. Perhaps for this reason, mothers—and fathers—of newborn infants might think twice about changing their hairstyles too often.

New sounds

Suddenly surrounded by an incredible variety of new sounds, your baby is able to pick up some familiar ones to aid him in his transition to outside life. In one recent study, investigators measured changes in babies' heart rates and respiration in response to hearing the voice of their mother versus the voice of another woman. The babies' heart rates decreased momentarily when they heard their mother's voice, showing that they were more attentive. This is an indication that the baby has formed a memory of the sound in the womb and is able to keep it until after birth.

Although a newborn's sense of hearing is further developed than his visual system and has had more practice before birth, his auditory system continues to become more sensitive and better adjusted to the sounds of his new environment. During his first few weeks, he continues to sleep soundly, even if his parents are moving about the room or talking to each other. At birth a baby's hearing is not as acute as ours, a situation that might not even be so bad at first, since babies have to get accustomed to so many new sounds. The muffled voices can be compared to the blurred picture they see at birth. Newborn babies hear about 15 to 20 dB less than adults hear, which is comparable to the effect of wearing a good pair of earplugs.

Dennis and Victoria Molfese, at Southern Illinois University, found out that newborn babies' brains already distinguish between speech and nonspeech sounds and that the two kinds of sound are processed in different hemispheres. To study this question, they placed a soft cap with little electrodes over the baby's scalp. This doesn't disturb the baby. When the room was completely quiet, the researchers measured the base level of electrical activity in the baby's brain. When the baby hears a sound, special waves of electrical activity called event related potentials (ERP) appear over the part of the auditory cortex specialized for processing sound.

When the investigators compared this wave to the baby's baseline activity, they found that the response to speech sounds was stronger in the left hemisphere and to nonspeech sounds stronger in the right.

Using tongue and nose

Like vision and hearing, a baby's senses of taste and smell act as guides in the world outside the uterus. Both senses help him find nourishment and form a comforting bond between him and his mother. A newborn baby is particularly responsive to the smell of the area around his mother's nipple. Within minutes after birth, this smell attracts the baby's attention and arouses him to actively seek the nipple with his mouth.

Exciting research in 1998 showed that babies have a memory for smells experienced in the uterus. Benoist Schaal and his colleagues at the Centre National de la Recherche Scientifique in Nouzilly, France, presented three-day-old babies with two pads, one infused with the odor of their own amniotic fluid and the other with the odor of amniotic fluid from another baby's mother. Voilà!—the babies preferred the smell of their own amniotic fluid.

Newborns are also quite sophisticated little tasters, distinctly revealing which taste they prefer—and this happens to be flavors that are sweet. The innate preference for sweetness may have had an evolutionary advantage, because it prompted our ancestors to seek out high energy fruits and berries.

When newborns find a taste inviting, their faces relax in an expression of contentment, and they may purse their lips or stick out the tip of their tongue. But when they try something that doesn't suit their taste, they pucker up their faces in disgust or turn their heads away. This is communication, a baby's very first dinner conversation. It is interesting that the baby's facial expressions of pleasure or disgust are built-in and that they are the same as those of adults. The complex sets of muscle movements needed for these expressions are mainly coordinated by a group of structures in the brain called the basal ganglia, which then send the signals to the brainstem.

A baby's facial expressions of pleasure or disgust are reflected in brain electrical activity. In one study, Nathan Fox and Richard J. Davidson observed infants' facial expressions and at the same time used electroencephalographic (EEG) techniques to measure infants' brain activity when the babies were given sweet and sour liquids to taste. When the babies tasted the sweet solution they showed greater activation in the left frontal hemisphere, and when they tasted lemon juice, they showed more activation on the right side.

Although newborns are capable of distinguishing odors and tastes and remembering them at least for a short period, this does not determine the child's tendencies to prefer a specific type of food later. Eating a balanced diet during pregnancy and nursing makes sense for both mother and child, but it will not "program" a baby to love garlic or broccoli later.

Getting the feel of things

In importance to human well-being, touch ranks just slightly below the need for oxygen, water, or food. It is by touch that your baby registers his first experiences with the world, when he is welcomed at birth by caring hands. It is by touch that he begins to shape his emotional ties to other members of his species. New parents automatically stretch out their hands to feel the miracle of their baby's perfectly shaped hands and feet and thrill at the texture of their baby's soft skin. You take your baby in your arms and press him close to your heart. He responds to your embrace and the warmth of your skin by snuggling closer and relaxing his whole body.

A baby's sense of touch is already well advanced at birth, thanks to his developing nervous system and his experiences with touch in the uterus, but everything in the uterus was smooth and warm. It's a whole new world out here to explore. To do this, a baby uses not only his hands, but also his mouth. The area around the mouth has a particularly large number of touch receptors, which got an early start in the womb. This is especially important right now for a baby's ability to locate his mother's nipple and find nourishment.

The sense of touch is also responsible for alerting a baby to potential dangers. Receptors in the baby's skin register changes in temperature and pressure, and if these changes exceed a certain level, they are felt as pain. Unfortunately, it was assumed as late as the middle of the twentieth century that the newborn infant had no sensation of pain, and if a newborn needed surgery, it was done without an anesthetic to avoid the risk of slowing the infant's breathing. However, we know today that the nerve tracts that carry pain sensations from the skin to the somatosensory area of the baby's cerebral cortex are already established around the twenty-sixth week after fertilization. For this reason, pain-reducing measures should be used wherever possible.

A study showed that two days after circumcision, newborn babies who had received a local anesthetic prior to the procedure were less irritable, showed better motor responses, and were better able to calm down than babies who had had no local anesthetic.

Muscles in Action

One of the first things you noticed when you tenderly placed your finger on the palm of your baby's wondrously shaped miniature hand is that the baby's fingers instantly and automatically closed around your finger in a tight grasp and didn't let go. The baby's grip is so firm that he can even be gently pulled to a sitting position. The baby's automatic grasp is called the palmar grasp reflex, one of a whole array of reflexes that babies come equipped with at birth. While some reflexes will be useful all through life, others, the so-called neonatal reflexes, will disappear over the course of the baby's first year as the cortex gradually takes control over the brainstem and movements become voluntary.

Reflexes can be a very efficient means of automatically doing what is best for us. When a bright light shines at us, our eyelids immediately close to protect our eyes. The signal goes only as far as the brainstem, not higher up to the cortex, where thinking takes place. If it did, we might not be able to act fast enough. The nerve pathways for some reflexes, such as pulling our hand away when we touch a hot stove, go through the spinal cord.

The palmar grasp reflex is the work of the brainstem circuit. Touch sensors in the skin of the baby's palm send an electric message over the spinal cord to the brainstem. The brainstem immediately sends a signal back over the spinal cord to the muscles of the fingers, telling them to grasp. A baby can't react differently. The touch sensation automatically leads to the baby's clenching of his fist. If this reflex did not disappear later, it would be impossible for the infant to explore the environment with his hands.

Several reflexes are essential for helping newborns find nourishment. When the baby's lips touch the warm skin of his mother's breast, he automatically makes eager movements of his mouth to find his mother's nipple. The "rooting" reflex is followed by the reflexes of sucking and swallowing. It all sounds very simple, but in reality some rather sophisticated coordination is involved, especially when feeding reflexes have to work together with breathing. This is the task of the brainstem, but even in healthy neonates, the coordinated actions need a few days of practice.

However, the actions of newborns are not only reflexes or the result of random, involuntary signals from the brainstem. Babies turn their heads and wave their arms and legs around, and it turns out that some of this activity is less random than it seems to us at first glance. One study showed that soon after birth, babies begin to use visual cues to help them guide their movements. When the babies were lying on their backs with their head turned to one side, they moved their arms into their field of vision so that they could see them. And when they could see their arms, they moved them around more.

George Butterworth and Brian Hopkins made some surprising discoveries when they observed a typical movement that babies make even before birth. They bring their hand to their mouth. At first this seemed to be a random movement, but then Butterworth and Hopkins noticed that when the newborn babies brought their hands toward their mouth, their mouth opened *before* the hand got there. The muscles of the baby's mouth must have received an advance signal telling them to get ready. The mouth's "anticipation" of the arrival of the hand is a very early hint of the baby's later ability to carry out purposeful movements.

The researchers also observed to their amazement that when the baby's *own* fingers touched his mouth, the infant did not automatically show the rooting reflex, which the baby usually shows when another person touches his cheek. Philippe Rochat confirmed these observations in 1998 and added that this might be a sign of the baby's emerging ability to perceive his own body as separate from another, the very beginning of a distinction between the self and nonself.

A Personalized Nervous System

That a baby's nervous system is tuned to individual levels of sensitivity can be seen in the baby's reactions to a routine procedure shortly after birth. A small blood sample is taken from the baby's heel. This is for the purpose of metabolic screening, a routine procedure carried out to determine if an infant is at risk for serious brain disorders if not treated soon. Babies startle at the prick of the lancet and begin to cry.

Reacting to pain is part of a system that is essential for preparing the body to respond to danger. In a threatening situation, or in response to intense pain, neurotransmitters and hormones are released into the bloodstream. The neurotransmitter norepinephrine, for example, puts the body into a state of readiness for action, while the stress hormone cortisol is important for keeping up the body's endurance. The basic components of this system are already beginning to function around the eighteenth week after fertilization.

Michael Lewis, at the University of Medicine and Dentistry of New Jersey, Robert Wood Johnson Medical School, compared

infants' individual reactions to the neonatal "heel-stick" procedure. He found two distinct groups of infants: the "high reactives," who cried instantly and loudly and took longer to quiet
down, and the "low reactives," who cried less intensely and
calmed down faster. He found that the high reactives had a higher
level of the stress hormone cortisol than the low reactives. This
was consistent with findings of another group, whose results
showed higher levels of cortisol in infants who reacted more intensely than others to the pain of circumcision.

Since the babies were only a few days old, the different reactions appeared to be due to inborn settings of the infants' nervous
systems. However, Lewis and his colleagues had to consider the
possibility that they were due to the infants' state at the moment.
Maybe the baby who reacted more intensely was merely hungry
or upset? Or the child who took the procedure more calmly was
simply too drowsy to react? So the researchers decided to compare the intensity of the babies' reactions to the heel-prick to that
of their reactions to routine inoculations two months later. The
high reactive infants also reacted more intensely to the inoculation, while most of the low reactives took the inoculation more
calmly. The strength of the baby's reaction appeared to be a trait,
or characteristic reaction, rather than a state, or a momentary
situation. And as we shall see, this tendency can persist as the
child grows.

Ready to Learn

The moment your baby is out in the world, he gets right down to
finding out how things work. One of his first discoveries is how to
find nourishment. At first he needs a bit of help from his mother
to find the nipple and fit it into his tiny mouth. After a few tries,
most babies settle right down into the routine of sucking and
swallowing. Some babies need a bit more practice.

Babies soon learn that their actions can get them somewhere.
In one study, babies less than three days old showed that they
could not only recognize their mother's voice but could also regulate their sucking pattern to summon her voice. To demonstrate

this, investigators made a recording of each baby's mother reading a few pages of Dr. Seuss. Before the test began, the babies were given pacifiers to suck on, and their baseline sucking rate was measured. Then earphones were put over their heads, and they were ready to begin. When a baby sucked faster than his baseline rate, he heard his mother's voice, and when he sucked slower, he heard another female voice. The babies quickly learned to suck faster so that they could hear their mother's voice.

Being able to form an association between one event and another is a basic form of learning. To see if newborns could do this, researchers made use of what babies do naturally when they taste something sweet: they begin to make sucking movements. This is something they don't automatically do when they are stroked on the forehead. So the investigators gently stroked babies on the forehead before feeding them a small amount of sugar water. They did this several times, until the babies began to make the sucking movements as soon as they were stroked on the forehead. The infants had made the association between being stroked on the forehead and tasting the sweet liquid.

The babies went a step further: they expected the sugar water to follow the stroking. When they were *not* given sugar water after being stroked on the forehead, they protested by frowning or crying. Their expectations were obviously not fulfilled. The ability to form associations leading to predictions of what will happen next is built into a baby's nervous system, and learning is off to a running start.

An Interest in People

After your baby's confinement in the uterus, he could well exclaim with Miranda in Shakespeare's *Tempest*: "O, brave new world, that has such people in't!" He may not yet be the life of the party, but he is already interacting with the people in his life. He responds to his father's touch, snuggles up to his mother's breast, and looks up trying to follow his parent's face when he is being talked to.

Babies have a special affinity for human sounds. When one baby starts crying in the nursery, several others are likely to chime in. Infants are able to distinguish between a real human cry and a synthetic imitation. One study showed that when newborns in the nursery heard a tape recording of infants crying and one of a computer simulation of babies crying, the infant listeners cried more often in response to the authentic crying tape.

Babies are even able to recognize their own voices. In 1999, Marco Dondi, Francesca Simion, and Giovanna Caltran, at the University of Padua in Italy, found that newborns are able to sense the difference between a recording of their own cry and that of another baby. When the babies heard the other voice, they stopped sucking on a pacifier, showing that they were more attentive. They were also more likely to squeeze their eyes closed and wrinkle their brows in a grimace. Their own familiar voice must have sounded more reassuring because when they heard it, they calmly went on sucking.

In addition to human sounds, human faces have a special attraction for newborn infants. Babies react differently to faces than to other visual patterns. In an illustration of this phenomenon, Mark H. Johnson and his collaborators at the MRC Cognitive Development Unit, in London, England, showed that newborn infants turn head and eyes significantly more toward a schematic drawing of a face (A), they move their eyes less often to a scrambled version of the drawing (B), and even less to an empty outline of a head shape (C). The babies were not only able to see differences in the patterns of the lines and shapes: they interpreted them. They responded to the features as a whole and not merely as a collection of separate details. And they preferred the face.

Nothing could be more fascinating to parents than to see their new baby watching them intently as they stretch or purse their lips and wiggle their tongue. Andrew Meltzoff, a psychologist at the University of Washington, observed what newborn babies do in response to these inviting expressions. In one set of investigations, the researchers demonstrated either a tongue protrusion or

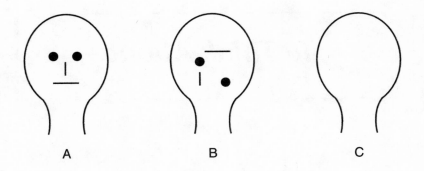

Looked at:

Most frequently　A. Face-shape with simple features

Less frequently　B. Face-shape with scrambled features

Least frequently　C. Outline of face-shape without features

Newborns prefer a schematic face to a head shape with scrambled features or an empty head shape. They show this by looking at the schematic face more than the other shapes. (Adapted from M. Johnson et al., 1991)

a round open mouth for 20 seconds and then assumed a passive-face pose for 20 seconds between showing the expressions. The experiment was videotaped and scored afterwards by an observer who was not shown what faces the adults were making. The results showed that the infants were more likely to copy the adult's expression.

Meltzoff considers this a form of early imitation, a mechanism that would facilitate social learning. Other researchers remain skeptical. They were either not able to replicate Meltzoff's results, or they suggest that the baby's responses are automatic and not really imitation. The question is still open. However, the fact remains: it's fun to make faces at babies and see what they do.

To Think About

Is birth traumatic?

Because birth requires both strenuous physical effort and the ability of the baby's body to adapt instantly to entirely different surroundings, people have sometimes referred to the normal birth process as a "traumatic" event and claimed that it could have consequences for later development. In 1929 Otto Rank, one of Freud's students, was even of the opinion (in contrast to Freud himself) that the trauma of birth was the main cause of neurotic anxiety in the adult. While birth is certainly a major challenge for the baby, Nature has equipped the newborn brain and body to handle this stress. There is no evidence that the baby remembers anything about his birth or that the experience of the birth process has any direct lasting effect later in life.

In medicine, the term *birth trauma* has another meaning: it refers to physical complications, such as a lack of oxygen, affecting the newborn during the neonatal period. Years ago birth complications used to occur more frequently. But thanks to the great progress in obstetrics and perinatal medicine, birth trauma is now a rather rare event. Birth trauma was once assumed to be the main cause of cerebral palsy (CP). However, in a study published in 1995, investigators found that only 14 percent of the children with CP had had a birth trauma. The vast majority of the children with CP suffered from prenatal complications such as intrauterine infection.

Since birth is indeed a complete transition, would it be an advantage to ease the infant gradually into his new environment—prepare a soft landing for him—by having the birth take place in water, for example? The idea is tempting, although it cannot really be called a "natural" proceeding for human beings. As far as the baby is concerned, no studies have proven that a water birth has any lasting positive or negative effect. However, mothers may find the warm water more relaxing, and this is a very important point.

What is "rooming-in" all about?

Thanks to modern maternity ward arrangements, it is now possible for a mother to have her baby in her room. This gives her a chance to become acquainted with the new baby and with his feeding and sleeping rhythms. Some research suggests that rooming-in facilities in maternity wards have helped mothers and new babies adapt better to nursing. But the new mother should feel free to leave the baby in the nursery when she wants a rest.

How important is body contact?

The question of body contact between mother (or father) and infant immediately after birth or during the next few days has been raised from time to time because people feared that an absence of early physical contact could have a lasting negative effect on future parent–child relationships. In a 1992 survey of studies on this topic, Diane E. Eyer showed that this is not the case in human beings. A close, caring relationship is not the result of a few moments of body contact or even a few weeks of being together. It unfolds over years through countless interactions, and the great majority of mothers and fathers form strong relationships to their infants under all kinds of circumstances, as parents of adopted children—and the children themselves—can testify.

For premature infants, a practice called kangaroo care is often recommended, in which the baby is held or carried next to the parent's skin. This fosters the contact between parents and their premature baby and provides natural warmth beneficial for babies whose nervous systems are not quite ready for the outside world.

Noise?

Newborn babies don't hear as well as you do, so there is no need to whisper or tiptoe through the room when they are sleeping. But since studies have shown that exposure to excessive noise is harmful to unborn infants, it is certainly a good idea to avoid rock concerts, loud stereo music in cars, and proximity to the roaring engines of jet planes.

Part Two
The First Year

Since babies are as different as flowers in a meadow, we have chosen to illustrate some of this great diversity by introducing a fictional cast of characters. Through "Emily" and her friends we get a small glimpse of the wide range of individual traits and possible ways in which children can develop.

To begin at the beginning, let's introduce Emily's parents, Deborah and Allen, a young professional couple living in a small university town. Their first child, Emily, was an "easy" baby. Two years after Emily was born, her brother, Andrew, arrived and turned out to be a challenge for his unsuspecting parents.

As you go through the book, you meet some of Emily's friends: Anna, Matthew, Sonja, Steven, and Tommy. When the group meets to celebrate Emily's birthday, we are able to follow the children's growing skills and their changing behavior in the group from infancy to their entrance into the world of school.

One Candle for Emily

Emily eagerly imitates her parents as they take a deep breath and exaggerate puffing out the one big candle on the top of her first birthday cake. She scoops up a generous slice with both hands and stuffs it into her mouth with delight. Then, with bits of frosting in her hair and crumbs still clinging to her collar, she wiggles out of her mother's arms and slithers down to the floor.

Deborah and Allen have invited several other neighborhood mothers and fathers to come with their infants to enjoy a lively afternoon. The babies crawl around independently like brightly colored ladybugs, tossing each other a glance from time to time. Although the children are roughly all the same age, they show a wide range of abilities and very individual sets of personal characteristics.

Anna cuddles close to her mother and glances warily at the approach of a new face. When Deborah smiles at her and attempts a friendly chat, Anna begins to breathe audibly, turns to hide her face in her mother's blouse, and cries.

Matthew toddles over to his mother to show her a puppet he

found under the coffee table. She can't help remarking that Matthew has been walking for a whole month already. Sonja persists in trying to gain possession of her mother's purse and explore its contents. Steven sits placidly off to one side, sucking his thumb and giving the chime ball a casual push from time to time. Rambunctious Tommy crawls rapidly off to grab one toy after another, only to toss them aside without further ado.

One of the children's numerous "expeditions" in the party room calls our attention to some of the remarkable capabilities of one-year-olds. Out of the kaleidoscopic array of scrambling infants, bright toys, and wrapping paper flung about in all directions, Emily spies her favorite bear slumped over on the sofa. Her "little wheels" begin to turn, and she forms a plan to get her bear.

Off she goes, hands and knees alternating in rhythmic fashion, eyes focused on her friend. Nothing can distract her. Alas, she has underestimated the height of the sofa, and her arm isn't quite long enough to reach the bear's paw. "Ma-ma," she calls out to her mother, but at that moment her mother is helping Anna's father to another cup of coffee. Emily doesn't give up. She grabs the end of the soft cushion and pulls herself up, reaches out, and draws Bear toward her. She glances over at her mother and grins triumphantly from ear to ear as she gives her bear a loving hug. She has reached her goal in spite of obstacles, and she is pleased with herself.

With all the other guests around, toddlers and babies and their happily chatting parents, Emily's mother hasn't a chance to catch her breath, to say nothing of having time to think back over how Emily has changed from the cuddly little bundle she brought home from the hospital a year ago. To appreciate the marvelous developments that take place during a baby's first year, let's begin with the adjustment phase of the first few months and then take a look at how a baby begins to explore her world.

Getting Started

When our son was born shortly before midnight in California, I proudly called my father-in-law in Massachusetts to tell him the news. His first question was, "And what are you going to do now, Norbert?" I was a bit puzzled by this but mumbled something about "going out with friends to celebrate." And then he said in his quiet New England manner, "No, go right home and get your last full night's sleep. First the baby will keep you awake, then you'll worry about where he's gone with the car, and then you'll have rheumatism." Things didn't turn out so badly, but it is a fact that the first few weeks of home life are a period of major adjustment for both babies and their families.

Just when you have finished nursing your baby and are desperately hoping to catch up on a few minutes of lost sleep, the baby starts to fret and cry for no apparent reason. Newborns are a bit like airline passengers who have just arrived at their destination after a long journey. The travelers may have experienced a tedious flight, the turmoil of a sudden change of climate, the

51

excitement of an exotic world of new sensations, and a major upheaval of their biological rhythms, but all this is trivial compared to what a newborn infant goes through. It is not at all surprising that it takes a baby a while to get settled and feel at home in her new surroundings.

Less Crying and Irritability

Your baby's nervous system still has to adjust to her new environment and to all the new sensations from inside and outside her body. Even before birth, the baby's cortex has begun to sort these sensations out and distinguish between meaningful and non-meaningful signals. However, for about the first three months, many brainstem circuits that involve internal organs are not yet controlled by the cerebral cortex, so they act reflexively. All kinds of signals are arriving in her brainstem, and the brainstem neurons do just what they are supposed to do—they automatically send them on. The baby's only way to respond is by crying and fussing. If she seems irritable and restless, it's not her fault. She has no other way to express the turmoil inside her. Fortunately, in the course of the next few months things will change for the better.

After about three months, you will breathe a sigh of relief. Around this time, babies show a clear reduction of "irritability" and crying. The average total crying time per day goes down from three hours to just one hour. This is a sign that the baby's cerebral cortex is gradually exerting more control over brainstem nerves that were sending the excitatory messages to the muscles. Now the connections between the cortex and the brainstem are becoming more efficient, making it possible for the cortex to block these random signals.

Rock-a-bye Baby

Babies love to be held and rocked softly back and forth. When you do this, they usually get very still, their eyelids slowly open and close, and in a few minutes they are most likely to fall asleep. A short tour "backstage" will show us more about why babies are comforted by gentle motion. Behind the scenes, a system essential for balance and control of body position is quietly carrying out its duties. If baby brain development were a film, the vestibular system would have a prominent place in the credits.

Unfortunately, we are most aware of the vestibular system when we suffer the effects of a choppy flight or when we spin around too fast on the dance floor. It coordinates the movements of our head, eyes, trunk, and limbs and allows us to keep our sense of balance. The main organ of the vestibular system is a structure called the labyrinth, located next to the inner ear.

The vestibular system is basically established and doing its job at birth. A newborn can lift her head momentarily in a prone position and turn her head to the side, allowing her to breathe freely. The baby turns her head and body to accommodate a change in position. Changing the baby's position can have an effect on general alertness. Holding the baby upright makes her more alert and attentive, and holding her horizontally and close to your body allows her to relax and become drowsy. We see the effects of body position on our own arousal level when we "sit up and take notice" or when we lie down to watch television.

The fact that the vestibular system is fully myelinated and functional in newborns may explain why gentle, horizontal, rocking movements are so effective in quieting babies.

A Welcome Relief

It may be small consolation for bleary-eyed parents to know that newborn infants sleep on average as much as 16 hours a day and are awake for only about 8 hours. Unfortunately, the 16 hours of sleep are distributed in various-sized chunks over the 24-hour period. Thus, parents of a new baby are often lucky to be able to sleep for 3 to 4 hours at a time. Sometimes it seems as if the baby has confused night with daytime or was born into the wrong time zone.

After about three months, the situation changes. Babies are now able to sleep for eight hours at a time, not bad for allowing parents to get a good night's sleep. By three months, 70 percent of babies sleep through the night, and by five months, 90 percent do. A baby's total nighttime sleep increases to around 12 hours by the end of the first year.

In addition to longer periods of sleep, the phases become more regular. The pattern of day–night cycles is called *circadian rhythm*, after the Latin words *circa* ("about") and *dies* ("day"), and the phenomenon comes as balm to the souls of distraught and sleepless parents. Circadian rhythm first appears when the baby is about one month old, but it becomes more distinct during the following two months.

Circadian rhythm is basically regulated by alternations of light and dark. In the morning, when light hits the retina at the back of the baby's eye, a signal goes to the hypothalamus, a brain structure that plays an important role in the regulation of sleep–wake cycles.

The hypothalamus relays the signal to the pineal gland, a tiny cone-shaped structure located in the center of the brain. The pineal gland slows down its production of melatonin, a substance involved in inducing sleep. When there is more light, less melatonin is produced, and in the evening, when the light is low, melatonin production increases. Before this system can function smoothly, some changes have to take place in the baby's body and nervous system. While she was in the uterus, her mother's melatonin entered her bloodstream through the umbilical cord. Now her own glands have to take over the production of substances that regulate sleep. She has to adjust to the contrast between daylight, or strong lighting, and darkness on her own—the price of autonomy.

Very little melatonin can be detected in the plasma of newborn infants, and they don't seem to have any pattern of melatonin secretion. But around the age of two to three months the baby's pineal gland matures and begins to secrete melatonin in a rhythmical pattern. As circadian rhythm becomes established, the baby begins to show regular cycles of sleep and waking states, with the night phase longer than the daytime nap periods. How easily a baby falls asleep and how much sleep she needs depends on the infant's individual nervous system, and—as in adults— there is a wide range of variability. Some babies are real "night owls," while others gladly turn in when the sun goes down.

A baby's body temperature fluctuates with circadian rhythm. During the period from birth to one month, before circadian rhythm has become established, the baby's temperature remains constant at about 98.1°F around the clock. At three months her temperature increases to a peak of about 99.3°F at around four in the afternoon. Thereafter, her temperature goes down to about 97.9°F at midnight and begins to rise again toward morning. In the following months, the night plateau in temperature becomes longer and lasts until about five o'clock in the morning. Adults whose temperature begins to rise early are the ones who get up bright and early and greet the world with a brisk "Good morning." Those whose temperature rise begins later are likely to be the ones who struggle out of bed wondering "what's good" about it.

Not only does the baby's sleep–wake cycle become more regular around the age of three to four months, the pattern of her sleep also changes. She shows distinct phases of both deep sleep and rapid eye movement (REM) sleep. REM sleep is a "lighter" kind of sleep, characterized by more general body activity and the rapid eye movements that give it its name. You can sometimes see the rapid movements of the baby's eyeball under her closed eyelids. During REM sleep is when adults are most likely to dream. Maybe babies do too. Less stimulation enters the brain from the outside world during sleep. Instead, the stimulation comes from within the brain itself. We can compare it to the internal (endogenous) stimulation in the baby's visual system before birth, when it was crucial for setting up basic connections.

Since stimulation is beneficial for the growth and specialization of neurons, we can speculate that REM sleep plays a role in the development of the baby's brain, especially during the first few months. Most of the newborn's sleeping time is spent in REM sleep. At three months, the amount decreases to about one-third of the total sleep time and at six months, to only one-fifth.

At the same time that REM sleep time is decreasing, the baby gradually stays awake for longer periods, and she is both more attentive and more capable of dealing with all the new impressions of her environment. Her brain now receives abundant stimulation from the outside world.

First Visit to the Doctor

Between two and four weeks after a baby is born, she usually makes her first visit to her doctor. Knowing what pediatricians are looking at when they examine a newborn baby can be reassuring and informative for new parents. Hands-on experience is even better. One study compared three groups of mothers. The first group heard only the results of the assessment procedures; the second group watched the examination; and a third group actually performed some of the tests with their own babies under the watchful eyes of the examiner. This gave the mothers a chance to see and feel for themselves how their baby responded. One month later, the investigators visited the mothers at home and observed how they interacted with their infants. The mothers who had actively participated in the test were more responsive to their babies than the mothers in the first two groups.

The newborn assessment procedure went something like this: the doctor checked the baby's ability to follow objects with her eyes and her reactions to sounds. The mothers were relieved to see for themselves that their babies could see and hear normally. Then the doctor checked the babies' muscle tone and reflexes. The mothers watched spellbound as the doctor held each infant around the waist and lowered her to the point where her bare feet touched the surface of the examination table. The babies put one foot in front of the other to make a kind of stepping movement. The doctor explained that after a few months, the babies wouldn't do this anymore. The newborn stepping reflex demonstrates a phenomenon that we saw in the chapter on life in the womb;

namely, that babies favor one side over the other. Most newborn infants start off with their right foot.

The doctor asked the mothers about their baby's behavior. How much did the baby sleep? How often was she hungry? How often was she fussy? How would the mothers describe their baby's "personality"? While the doctor was talking with the mother, some babies began to fuss or cry. The doctor then explained how even newborns find ways of calming themselves when in distress. Many have already discovered comforting "strategies," such as sucking their thumb. The rhythmic sucking lowers the baby's arousal level and quiets the movements of her limbs.

If the baby didn't calm down by herself, the mother could try speaking softly into one of the baby's ears to see if this would attract the baby's attention and help her forget her momentary distress. Or the mother could hold both of the baby's hands and keep them still. Restricting the movements of the limbs reduces excess stimulation caused by their activity. This observation has led in some cultures to the practice known as swaddling. When I was a baby, my own pediatrician advised my mother to gently stroke my arms to quiet me down. According to her, the method was most effective—and it still is.

How Different Babies Can Be!

When their second child, Andrew, was born, Deborah would often look back on the first few months with Baby Emily as if it had been a marvelous dream. She and Allen had been the envy of all the other young parents in their circle. Bright-eyed Emily was an energetic bundle who took to nursing like the proverbial duck to water. A few weeks after they took her home from the hospital, she slept right through the night. Sometimes Deborah's mother came over to look after her, and Emily placidly let her bounce her around or change her diapers without any sign of annoyance. When Deborah's mother wasn't available for baby-sitting, the young parents proudly took Emily out to dinner with them in her infant carrier. They put her down beside them, and she either looked around quietly or drifted off to sleep. The sound of dishes rattling off to the kitchen or the lights above her didn't disturb her rest. She had a sunny, easy disposition and a ready smile. In short, she was her parents' idea of the perfect baby.

When their son Andrew was born two years later, Allen and Deborah were completely baffled. Andrew was the rabble-rouser in the nursery, screamed unmercifully, and thrashed his arms and legs in the air as if he wanted to push the whole world away from him. And when he got home, his behavior didn't change. Andrew tired easily when he was nursing, so he never drank enough at any one time and consequently was hungry all the time. The circles around the eyes of Allen and Deborah grew larger and darker, while their patience shrank in proportion. Andrew got upset when Deborah's well-meaning mother picked him up. Gone were the romantic candlelight dinners. Andrew wasn't even happy to be taken out on walks! He woke up and cried when he heard a noise or when a light went on nearby. He had bouts of colic, and his bottom was red with diaper rash. He was fussy and demanding most of the time. In retrospect, Emily had been a "little angel," a "sunbeam." Andrew was more like a finicky hotel guest. How could two children in the same family be so different? More worrying still, would Andrew always be this way?

Each child has an individual manner of responding to what she experiences and a personal pair of mental "colored glasses" with which she views the world. Emily is easygoing and sociable, more apt to take changes in stride. Andrew, on the other hand, reacts negatively to any tampering with his routine. This personal set of basic characteristics is called *temperament*. Some of the traits belonging to temperament remain detectable all through life. However, they are constantly molded over time by heredity, maturation, and experience.

Jerome Kagan and his colleagues at the Harvard Infant Study are conducting a long-term study of infant temperament using direct observation of the children's behavior. They chose to concentrate on one characteristic of temperament: how a child reacts to novelty. By focusing on a standard set of behaviors to observe and record all children they study, they have been able to compare children's reactions as they get older. The researchers have tailored their experiments to the age of the children.

They test the infants' responses at the age of four months to an unusual sound and to a mobile waved close to the infant's face. At 14 and 20 months the investigators observe the children's degree of fearfulness in unfamiliar situations, at three years their sociability with a new child, and at seven years their sociability in a group of children.

When the babies in the study were four months old, they were presented with some situations that were harmless but nonetheless a bit discomforting to infants of that age. As an example, let's pretend that Emily and her friend Anna were two of the babies in these studies. In the first situation, Emily was propped up in an infant seat on a table, while her mother remained in a corner of the room. The investigator then played a taped, distorted female voice telling the child that she had been a "very good baby." Emily sat quietly and unperturbed, only turning her head slightly and slowly moving her big, bright eyes to one side. In contrast, Emily's friend Anna showed distinct signs of distress when she heard the strange voice. She writhed uneasily and arched her back, made stiff movements with her arms and legs as if to push herself away, and pulled the corners of her mouth down to make a very sad face. She began to cry.

In another situation, a bright children's mobile was set in motion close to the baby's head. Anna showed the same reaction that she showed at the unfamiliar voice, while Emily remained calm. The study found that 20 percent of the babies were like Anna and showed signs of being very disturbed (high reacting), while 40 percent of the babies were more like Emily and took it calmly (low reacting). The researchers also noted that the "Emilys" were more likely than the "Annas" to be sociable and to smile frequently.

The investigators in the Harvard Infant Study concentrated on the two extremes, the children who could be most clearly categorized as high reactives, the "Annas," or as low reactives, the

"Emilys." Like the differences in the newborns' reactions to the heel-stick procedure and the two-month-olds' reactions to the inoculation, the variety of ways in which four-month-old children react to unfamiliar events have a biological basis. Their nervous systems are not all preset the same way. Anna's nervous system is more sensitive and reacts more strongly to disturbances than Emily's.

The "settings" of a baby's nervous system are the result of genetic factors and conditions in the uterus. It is possible that prolonged, severe stress of the mother may make them more sensitive. Being aware of the individual set points of their baby's nervous system and how these partly explain why one child cries or moves more than another or has more difficulty calming down is a boon to parents. Knowing that babies have different nervous systems may make you more patient and give you more confidence.

However, the temperament characteristics seen in young infants are subject to change with time. By the time she enters school, Anna will most likely be less upset by novelty than she was as an infant. Her nervous system will mature, and at the same time, her experiences in her world will affect the way she reacts to the challenge of unfamiliarity.

Temperament and Parenting

Knowing about a baby's temperament and learning ways to handle babies that are more irritable or "difficult" than others can have important consequences for the way parents interact with their child. Dymphna C. van den Boom, of Leiden University in the Netherlands, found that mothers tended over time to increasingly ignore their difficult infants and to play less with them. She proposed that if the mothers learned other ways of handling their infants, they would become more responsive, and the babies, in turn, less irritable. This would improve the nature of the mother–child relationship and make it more enjoyable.

So van den Boom conducted a study in which mothers in the experimental group were given practical instructions in their homes on how to respond to their infants. The investigators selected infants that had high irritability scores on a standard neonatal behavioral assessment scale on the tenth and fifteenth days after birth. When the babies were six months old, investigators on the research team visited the infants in their homes and noted how the mothers interacted with their babies. For example, they noted what the mother did when the baby cooed or cried.

In several sessions over the following three months, the investigators showed the mothers in the experimental groups how they could pay attention to their infants' cues and respond appropriately. They encouraged the mothers to repeat the baby's sounds and to maintain silence when the infant's gaze was averted. They also helped the mothers find the best way to soothe their babies. For example, some babies like close physical contact, while others don't.

When the babies were nine months old, the investigators compared the mother–infant pairs in the experimental groups to those in the control groups that had had no instruction. They found that the mothers in the training groups were now more responsive, more stimulating, and more visually attentive to their infants. And their babies were more sociable, better able to soothe themselves, and cried less. In addition, the babies showed more interest in exploring their environment.

Doreen Arcus, of the Harvard Infant Study, showed that parenting style can have an effect on a child's fear of novel situations. The infants in her study were those who had been classed as either extremely high reactive or extremely low reactive on the basis of their behavior in unfamiliar situations at the age of four months. The investigators had found that the high reactive infants have a greater chance of being more fearful of unfamiliar persons, objects, and events at 14 months than the low reactive children. In a typical test situation, a stranger wearing a gas mask and a white coat approached the child. One of the objects was a rotating wire drum spun at varying speeds while noise-provoking objects were added. In another situation, a metal robot with flashing lights and a male voice appeared, and the child was invited to approach.

Arcus observed the infants at home from the age of 5 to 13 months and made note of both the baby's and the mother's behavior. She distinguished two main styles of maternal parenting: permissive and authoritative. In contrast to the adjective *authoritarian*, used for parents who demand absolute obedience and who may even inflict harsh punishment in order to achieve it, the term *authoritative* used here refers to parents who set and enforce guidelines in a way that is appropriate for the age and temperament of the child. The permissive mothers in the study tended to adapt to the baby and try to keep her happy. The authoritative mothers expected that the baby could not always be happy and they would have to help the baby learn to find means of coping on her own.

Early in the first year, the permissive mothers were more apt to respond immediately to the baby's crying and fretting. They thought all the baby's needs should be immediately fulfilled. The authoritative mothers first waited to give the baby a chance to calm down on her own. Later in the first year the permissive mothers gently guided the baby away from danger but did not otherwise interfere with the baby's explorations and tried to minimize frustrations. The authoritative mothers tended to be firmer and did not hesitate to set limits by saying no to behavior that was potentially dangerous or socially less acceptable.

At 14 months, the investigators observed the babies' reactions to unfamiliar situations, such as the appearance of a clown. While the mothers' parenting style had little or no impact on the low reacting infants, it did have an effect on the high reactors. Surprisingly, infants who had been classified as high reacting at 4 months and had permissive mothers showed even more anxiety than expected at 14 months. But the high reacting infants of the authoritative mothers were less fearful than expected. A possible reason for this could be that the authoritative mothers gave their infants more opportunity to practice coping with frustrations. In addition, the mother's calm, firm setting of limits may be perceived by the child as a confirmation of the mother's presence and support. The child may thus feel more secure in unfamiliar situations.

To Think About

What is the "right way" to care for babies?

There is no one right way that is good for everybody. A baby's
biological and emotional needs can be met in a great variety of
ways. Parenting practices depend to a great extent on the conditions and expectations of the culture the child is born into. In
some societies, women carry their infants around with them as
they go about their work in the fields. If you are running a large
corporation, directing traffic, or working in the hospital emergency room, this will most likely not be an option for you. Although many non-Western mothers carry their infants around
with them, they do not kiss and cuddle them or speak with them
as much as Western mothers do.

Child care practices also change with the times. During the
eighteenth century it was not unusual for affluent mothers to give
their children to another woman to nurse. Before the invention of
disposable diapers, early toilet training had obvious practical ad-
vantages.

Mothers and fathers should take time together to discuss the
arrangements for the new family member. Use common sense
and speak to your doctor, nurse, or midwife if you are not sure
what to do.

Breast or bottle?

Breast milk is Nature's own formula and provides the newcomer
with special protection from infections. Patience is necessary,

since both mother and baby may need a bit of practice. As Valerie Frankel says in the November 1999 issue of *Parenting*, "Nursing may be natural, but it's not always second nature." It may take about a week to establish a routine.

If breast-feeding is not possible, mothers should not worry that feeding their baby a carefully prepared formula from a bottle will have a negative effect on brain development. An advantage of bottle-feeding is that it allows fathers to have the pleasure of feeding the baby and sharing a bit of close time together.

How shall I handle crying and fussing during the baby's time of adjustment?

Knowing how a baby's nervous system develops can help you make your decisions about how to get through this often trying phase. In general, crying increases up to about six weeks and then diminishes over the next two to three months. Some babies cry more than others, no matter what their well-meaning parents do.

As you get to know your baby, you will become adept at "translating" many of her cries as hunger, tiredness, pain, or discomfort from a wet diaper or stomach gas. But sometimes the baby cries or seems restless for no apparent reason. These periods are more frequent toward evening. This is most likely due to the fact that the baby's brain is not yet ready to handle all the signals it's getting from the baby's internal organs. The baby's body is not yet producing melatonin, the brain chemical that helps regulate sleep on a regular schedule, so her sleep–wake patterns may be erratic.

Once a baby's nervous system has had time to get adjusted to her new surroundings, she settles into her own pattern of sleep–wake cycles and feeding times (rhythmicity). By five months, most babies sleep through the night. Feeding times also become more regular. Being aware of a baby's biological rhythms means that parents can anticipate the times when a baby will be hungry or restless. It is better to feed her, comfort her, or settle her down for a nap before she gets really upset. If a child's "schedule" appears random, with no clear pattern detectable, parents should guide the baby into a more consistent pattern, trying to set aside regular times for feeding, sleeping, and activity.

Because a baby's brain is very busy getting used to the multitude of new sensations around her, she will find it easier to calm down when there is less stimulation. Holding her firmly and rocking her gently is better than playing loud music or trying to distract her by dangling a toy in front of her eyes. On no account should you shake the baby. Her neck muscles are not yet strong enough to keep her head steady and prevent her brain from coming in contact with the hard skull. Baby shake trauma can be fatal or lead to serious brain injury. Seek help early if you are beginning to feel you cannot cope with your baby's constant crying.

Although parents can not eliminate bouts of crying, they can shorten their duration by carrying infants around at regular times distributed throughout the day and by following a routine in daily life, including mealtimes, sleep, and activity, such as play or going for a walk.

What about us?

This is a challenging time for you and your partner. You need time to be together to reinforce your own relationship and gather strength for your new responsibilities. Make arrangements to go out together once in a while without the children. If you are lucky to have parents or good friends who can take over for a while, you might even plan a week's vacation for just the two of you. Mothers (or fathers) who take care of children all day should have at least one free morning or afternoon per week to relax "off duty."

How can I help my highly reactive child now so she will be less fearful as a toddler?

For the child with a tendency to react strongly to any disturbance or change in routine, Doreen Arcus's study showed that a warm, authoritative parenting style is particularly beneficial. During the early months, you can help by establishing a stable routine in daily life. As the baby becomes more mobile, she needs the security of firm limits. For example, when she tries to climb up onto

the table, instead of just blocking her path without saying anything, look at her and say firmly, "No." If she continues to try to
climb up, say no again and calmly move her to a safer area of the
kitchen. Give her a pan and a wooden spoon if it's necessary to
distract her.

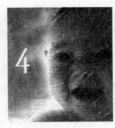

Exploring

Ababy's senses are his instruments of navigation, his antennas for taking in the abundance of new signals and giving them meaning. However, right from the very beginning, our little Columbus has been guiding his own vessel through uncharted waters and been fast at work putting his own map together.

During the course of the first year, a baby's view of the world expands with incredible speed. By the time he is three to four months old, his total waking time has gone up from 8 hours a day to a full 10 to 12 hours. He gradually stays awake for longer periods at a time. Spending less time in sleep means that he has more time to take in new impressions and experience new situations.

Being able to grasp an object and turn it over, to feel the touch of his mother's hands, the roughness of his father's beard, the texture of his soft bunny, adds sensations to be savored and incorporated into your baby's view of the world. New perspectives achieved by sitting, crawling, standing, and taking the first big step bring a wave of new impressions within his reach. Reflexes come to play a lesser role in his overall motor activity as he becomes capable of guiding his muscles with greater precision, using visual, auditory, and tactile information.

Parallel to his progress in using his basic tools, a baby gradually takes over more and more of the initiative. He not only sees. He looks. He not only hears. He listens. He can focus his attention on an attractive toy, form a plan of action, and carry it out.

A World of New Impressions

"Attention, please," is what all the new sensory impressions seem to be saying to your baby. His first few months are a time of rapid development in his sensory systems, allowing him to take in more detail, more colors, to distinguish sounds and melodies, to feel differences in texture with greater precision. The flood of new sensory impressions stimulates a special part of his brainstem, the reticular formation, ordering it to send out more neurotransmitters, substances that make him more alert, more on the lookout for exciting discoveries. This is similar to the way his nervous system was suddenly wakened at birth. He is now better able to detect objects in his environment, to focus on them, and to ignore the multitude of distracting and interfering impressions surrounding him. A baby can now remain calmly attentive to things he sees or hears for two minutes or more.

The expanding picture
Your two-month-old probably loves to sit on the table in a reclining infant seat and watch you eat supper. Around this time, babies are suddenly absolutely fascinated by all that is going on around them. They can now use their eyes much better, thanks to changes taking place in the brain, particularly in the visual system. During the first two months after birth, a baby's visual repertoire is limited. He can only see things about 1 foot away from his eyes. His eyes follow strong patterns reflexively. The ability to do this is the work of a special subcortical structure, the superior colliculus, located in the midbrain, above the brainstem. The

superior colliculus is involved in the detection of movements of objects in the visual field. It integrates information from the senses, making it possible to locate an object and to follow it. At first, a baby is more attracted by moving objects because his system for detecting motion is better developed than the pathway for detecting form and color.

By around two months, a baby begins to follow objects more smoothly because his cortex allows him more control. You may be surprised that for a short time, somewhere between two and four months, his gaze will be quite literally "captured" by what he sees directly in front of him. He can't seem to let go and shift his gaze to an object on the periphery of his field of vision. This momentary difficulty is a sign that the new system is taking over. Until now his looking movements were mainly reflexive. As his cortex gains control, he becomes able to direct his eyes at what he wants to see. At around three months, the axons of the nerves connecting the baby's cortex and his eyes are strengthened by their coat of myelin.

A baby can now recognize faces better because he sees not only contours but also the features within the shape of a face. The cones, the cells in his retina that are responsible for sharp vision and color, increase in number and begin to mature, making it possible for him to see objects more clearly. His visual acuity will continue to improve rapidly over the next two years and then more slowly until puberty. Because he can see more color, bright colors become increasingly attractive. When he is about three months old, he will enjoy that cute mobile your friends gave you as a present for the new baby.

In order to see one of the toys bouncing up and down on the strings, a baby has to focus his attention on it and see it as a separate item. The toy has to stand out as a whole from all the competing impressions in the background. At around four months, babies become better able to see single objects standing out from their surroundings.

G. Csibra and colleagues made an interesting observation when they showed infants a figure called a Kanizsa square. If you

Kanizsa square   Random pattern

(Test)   (Control)

"Kanizsa square" test: Looking at an illusory square and looking at a misaligned figure that doesn't cause the illusion lead to different electrical patterns in the brain.

look at the figure, you will see that the square is actually an illusion. It is not really there, but your brain puts it together so that you "see" it. When adults look at these pictures, a burst of high-frequency electric waves (gamma waves) can be observed in the higher visual areas of the brain.

The investigators presented infants with Kanizsa squares or with a control stimulus consisting of misaligned elements that do not produce the illusion of a square. The eight-month-olds, but not the six-month-olds, showed gamma activity like the adults when they viewed the virtual figures but not when they were looking at the control figures. The results suggest that maturational changes occurring in the brain around seven to eight months lead to this electroencephalographic (EEG) response. The findings also show that at this early age we can observe the brain at work.

At around four to seven months a baby becomes capable of perceiving objects in three dimensions. This is the work of the association areas of his visual cortex, which are undergoing intensive development around this time. These areas integrate the information from both eyes, which makes three-dimensional viewing possible. You may remember the critical wiring phase that took place before the baby was born. Then the necessary stimulation came from within. Now, to complete the link between the thalamus and visual cortex, the brain needs the stimulation provided by viewing experience in the outside world.

The impact of this sudden profusion of new visual sensations is tremendous. Between birth and eight months, the number of synapses in a baby's visual cortex increases 10-fold. It is now the highest it will ever be in his life—twice as many as in his adult visual cortex. This phase is called *blooming*, analogous to the explosion of blossoms in a summer garden. At around three to four months, the volume of the baby's visual cortex is the same as it will be when he is an adult, while his whole brain is only 50 percent of its adult size.

Blooming is followed by pruning. A baby's visual experience leads to a reduction in the number of synapses. Synapses that are frequently used become more stable, while unused synapses are cleared, or "pruned" away. From the time an infant is about eight months old, the number of synapses in his visual cortex begins to decrease to reach adult level around his 10th birthday. During this whole period, his visual system is being fine-tuned to process information with greater detail and precision. Maturation of the visual cortex is achieved just before puberty.

The baby's visual system can expect to find everything it needs for development in the baby's environment. This is called *experience-expectant activity*, and under normal conditions babies do not require any special visual training. However, if the baby cannot see adequately, his visual system will not develop properly. Therefore, it is crucial that visual defects such as cataracts be corrected as soon as possible. In a common condition, crossed eyes, infants see double when they look with both eyes. If this condition is not corrected, their brain gets used to using just one eye, and normal vision will not be possible later. If you think your child has crossed eyes, take him to a specialist. Treatment should be undertaken as soon as possible, while development is still going on in the child's visual system.

From sound to music

Most likely you just naturally start singing or humming softly to comfort your baby when he is upset and having trouble settling down to sleep. Music has a powerful impact on babies, just as it does on all of us. As the author Romain Rolland put it, "Music is the key to the soul." All over the world babies are comforted by lullabies, which usually have a slow, steady rhythm and a simple repetitive melody.

At around two to three months, your baby's interest in sounds really picks up. He may perk up and turn his head to listen to the tinkling of his Swiss music box or to his father whistling "Jingle Bells." Babies soon become rather discriminating listeners. An interesting study by Marcel Zentner and Jerome Kagan showed that infants as young as four months old are sensitive to harmony. The babies first heard a harmonious version of a pleasant children's melody. They calmly looked around the room and generally seemed pleased with the sound. But when they heard a discordant version of the same melody, they showed the infant equivalent of critics' jeers. They wrinkled up their faces in disgust and began to cry, as if the sounds were actually painful. Studies with adults have shown that dissonance activates areas of the limbic system involved in the processing of unpleasant sensations.

The baby's brain puts the single tones of his mother's lullaby together so they can be perceived as music. At birth, his cochlea, or inner ear, and his cochlear nerve are already practically mature, meaning that they can translate the air vibrations into electrical signals and transmit them to his brainstem. From his brainstem, the signals go to the thalamus and on to the auditory cortex. When you sing a few bars of "Rock-a-bye baby," the sound waves are translated into separate sensations for volume, pitch, and location. Only when a special area, the auditory association cortex, puts them together do they become the familiar notes of the melody.

A very interesting phenomenon takes place during a baby's first few months. While newborn babies are quite adept at turning their heads toward a sound, there is a short period, somewhere between two to four months, when their ability to do so is reduced for a while before it returns to its former level. This may be because a new system is taking over in the baby's brain. At first the baby's head turning is a reflex initiated by the brainstem. After the first few months, the cerebral cortex takes over control, which will make it possible for him to turn his head voluntarily. This U-shaped course of development is similar to what we saw taking place in the baby's visual system.

A lot of activity takes place in the auditory pathway during the first year. At around three months, a baby has more synapses in his auditory cortex than at any other time in his life. As in his visual cortex, their number will then be reduced and the remaining connections made more efficient. Between 6 and 10 months, the acoustic radiation, the connection between his thalamus and auditory cortex, is fully myelinated, indicating that it is functionally mature. However, parts of the auditory pathway are still under construction until he is about 10 years old. During this time, his ability to perceive and distinguish sounds improves.

Babies do not need extra stimulation in order to develop their sense of hearing. As we have seen with vision, the auditory system is tuned by normal experience with the great variety of sounds present in the environment. So it is essential that no hearing defects prevent or distort an infant's experience with sounds. The earlier a hearing defect is detected and treated, the better the chances that a child can compensate for it. Therefore, it should be diagnosed in the first months of life. If the baby has some hearing ability, a hearing aid can be highly effective. If the baby cannot hear at all, a device called a *cochlear implant*—to replace the inner ear—can be considered. More than 80 percent of the children who receive such an implant learn to speak and to understand language.

Good hearing is extremely important, not only because babies are learning so much about their physical world, but also because they are learning to interact and communicate with others. This is the time when the foundations of the baby's language and emotional systems are being laid down.

The senses work together

An infant's opportunities to explore and learn about his world are vastly multiplied by the fact that his sensory systems work not only as separate units but also pool their resources. We have seen that at birth the brainstem is already pretty good at coordinating the multimodal activity, but now the baby's cerebral cortex is getting into the game. In the cortex, information from the separate sensory pathways converges. His "updated" sensory systems are more efficiently connected and develop links to areas that are responsible for adding emotional color and storing memories.

Babies can combine visual and auditory information. In this they are like lip-readers. Patricia Kuhl and Andrew Meltzoff showed five-month-old babies filmed sequences of an adult making the sound *ee* as in "peep" or the *o* as in "pop." But in some of the sequences, the sound tracks had been switched. Two-thirds of the babies looked longer at the face that matched the sound, indicating that they recognized that the lip movement matched the sound they heard.

Other studies showed that babies link the senses of touch and vision. Since infants' mouths have an especially well-developed sense of touch, investigators gave one-month-old babies pacifiers to suck on. One group received a smooth, round pacifier, while the other group got a pacifier with little nubs. The babies sucked on the pacifiers for 90 seconds before the researchers removed them from the babies' mouths. The babies were then shown for the first time a choice of two models of the pacifiers made out of orange Styrofoam. To the researchers' surprise, the infants looked longer at the form they had been sucking on. Somehow the babies must have been able to register the smooth or nubbly surface of the pacifier. They compared this sensation with the picture in front of them and formed a temporary association between their tactile and their visual impressions, quite an accomplishment for a one-month-old baby.

A baby's senses are further linked together with his motor system. Information about what he sees, hears, and touches is put together by his brain and used to guide the movements of his muscles. By five months, when he hears the sound of the rattle, he can reach out and grasp it.

"Make Way for Baby!"

Matthew surprised his proud parents by taking his first steps when he was about 10 months old. He bravely took his hand from the arm of his father's reading chair and tottered off across the room, when a cheer from his proud parents was enough to throw him off balance. He collapsed in a somewhat puzzled heap on the floor and looked up bewildered: "What went wrong?" It was clear that some systems still needed practice in working together.

Shortly after the baby's birth, we already see the very first inklings of his ability to guide his muscles. Basic circuits are established that make possible not only reflex movements but also some spontaneous movements. However, learning to use muscles needs a lot of time and practice.

A baby's ability to lift and move his head may seem to us rather modest to be called a milestone, but it means that he can use his muscles to change his position, letting him take in more of his surroundings. At around two months, his control of his head and neck movements allows a range of vision of only about 90 degrees. However, between the second and fifth months this widens to about 180 degrees, meaning that he is now capable of looking around and following the movements of persons in the room.

At the same time that the baby's muscles are becoming stronger, the axons connecting his motor cortex to the neurons in his spinal cord are becoming more efficient. In the beginning, most of the connections within a baby's nervous system are like a bumpy, unpaved country road, over which signals travel slowly like stagecoaches. So Nature has found a clever way to pave the highway. Axons of the nerve cells are surrounded by layers of myelin, which insulate one axon from another and make the electric impulse go up to 20 times faster along the nerve. To reach this speed without myelin, the axons would have to be much thicker, and our spinal cord would be the size of a huge tree trunk about 3 yards in diameter. Picture yourself in front of the mirror!

Evidence that the connections between the baby's motor cortex and other parts of his motor system are becoming more efficiently linked is the disappearance of some of the neonatal reflexes. When Matthew was born, his fist closed automatically around his father's finger. This is a reflex initiated by the brainstem. Sometime when a baby is around three to five months old, the reflex grasp disappears, freeing him to manipulate objects and to explore the environment with his hands.

The baby's motor cortex is now better able to block the commands from his brainstem. The motor cortex is extremely versatile. It not only sends commands to activate muscles; it also has the important responsibility of blocking, or inhibiting, their activity. A balance between excitation and inhibition is crucial. Unchecked excitation could result in completely erratic movements, while too much inhibition could lead to a severe reduction of activity.

Between three and five months of age, the connections linking the baby's cortex and brainstem are undergoing a rapid phase of myelination. As we have seen in Chapter 3, this is also the time when fussing and bouts of crying for no apparent reason become less frequent.

When Matthew was four months old and his parents waved a bright red rattle in front of his eyes, he eagerly tried to take it in his hands. At first he made vague, swiping movements, but he learned very quickly to adjust the position of his arm and hand and to close his fist at just the right moment. By around five months, he could position his fingers in anticipation of the grasp, making it more likely to succeed.

A baby's seemingly simple reach requires the cooperation of many different areas in the brain, all of which are in a phase of intense development. With the help of the visual cortex, he sees the rattle. His motor cortex sends the appropriate commands down his spinal cord to his muscles, telling them what to do. Meanwhile, the cerebellum, located at the back of the brain, acts like an air traffic control tower. It receives information from the muscles, joints, and vestibular system and compares the actual movements with what the baby intended. The cerebellum corrects the commands to the muscles if they get off course. The cerebellum is important for learning how to use muscles. The connections between the cerebellum and the motor cortex develop rapidly at around three to four months. During the first year, the cerebellum is in a phase of rapid growth.

A baby's successful grasp of the rattle usually leads to putting it directly into his mouth. The pathway for touch sensations from this area is particularly well developed, making it especially inviting to explore new objects by mouthing them.

Getting from here to there

At five to seven months, babies can sit without support. Now both arms are free to use. Infants' muscles are stronger, and the cerebellum and vestibular system keep them from toppling over when they reach out to grab a toy. They can devote their attention to fingering and mouthing all the things lying around them. But soon this is not enough. They are off on their own.

The achievement of self-locomotion marks the beginning of a new era, and, like their possession of a new driver's license later, the end of a more restful period for you. Although babies start making the first movements that resemble crawling at around four weeks, it isn't until 7 to 10 months that they really take off. However, crawling isn't a completely standardized procedure; babies find individual crawling styles. Some babies even dispense with crawling altogether.

When he was nine months old, Matthew discovered that he could pull himself up to a standing position. He could even stand freely for a few seconds without holding onto anything. Holding onto the furniture or railings and pulling himself along became a much better way to get around than crawling, and soon he took his first real step—as Neil Armstrong might say, a small step for mankind but a giant leap for Matthew.

As a young doctor, I was fascinated by how babies learn so

much so fast. At that time, we often referred to a book by Arnold Gesell and Catherine Amatruda called *Developmental Diagnosis: Normal and Abnormal Child Development*, which contained neat line drawings of what children were supposed to be able to do at a certain age. So when I was a parent myself, I could hardly wait to follow our son's progress. I enthusiastically brought out my old black-bound copy of Dr. Gesell's classic and showed it to Elinore. When I began making remarks that our son was a bit slow on some of the points, she told me in no uncertain terms to put the book away. A few months later, when I couldn't find my book, I secretly suspected that she had something to do with its disappearance. I felt a bit sheepish when, 20 years later, I found it while I was cleaning out my desk.

We really shouldn't always compare our children to the other children on the block. But we do. The story of Steven and Matthew, based on the experience of a mother who came to our outpatient department, is typical.

When Steven was 11 months old, his mother invited a new friend and the friend's one-year-old son Matthew over for the afternoon. Steven was playing quietly by himself in his playpen when Matthew arrived. Steven's mother had to catch her breath when she saw how Matthew ran into the room, placed both hands on the railing of Steven's playpen, and vaulted elegantly over the top to land inside. In talking to her friend, Steven's mother learned that Matthew had been sitting on his own since he was 4 months old and walking freely since he was 10 months old. She began to worry that Steven might not ever catch up. To make matters worse, Steven's father kept comparing his son's progress to the scales in a textbook on infant development. Steven's mother was reduced to tears.

Her tears turned out to be entirely unnecessary, since Steven did indeed begin to walk by the time he was about 13 months old. Children show a wide range of variation in the time they begin to sit, crawl, and walk. This isn't surprising when we think of the many systems involved. Each structure develops very individually, as do the connections among them. In spite of the wide variation in the times children reach motor milestones, they all reach them in the same order. Although certain skills do come more naturally to some children than to others, it is important to remember that these often change as children grow up.

Memory and Learning

Here's a little game you can play with your baby when he is about four months old. When he is seated comfortably in his infant seat, show him an attractive toy, such as a small red ball. He will probably gaze at it intently. Show him the ball a few more times, and after a while he will not look at it any more. He's beginning to find this boring. After about 10 seconds, show him the red ball and at the same time, another equally bright and attractive toy; for example, a blue ball. He will most likely look right away at the *new* toy. He has compared the two, recognized the ball he has seen before, and now wants to take a closer look at the new toy.

This is an illustration of recognition memory, or the ability to distinguish between what the baby has seen before and what is new, and it is a crucial component of learning. To be able to recognize that something is different means that the baby has to be able to keep a memory of what was seen before. Recognition memory depends on the activity of the hippocampus. The greatest velocity of growth in the human hippocampus takes place between two and three months. The nerve cells there are rapidly becoming specialized for their particular functions, and sprouting a profusion of branches.

The novelty wave

The ability to distinguish between new and familiar is not limited to visual impressions. Babies can also distinguish sounds, tastes, and tactile impressions. This can now be seen directly by measuring the electric activity in the brain using EEG techniques. Ghislaine Dehaene-Lambertz recorded event related potentials (ERPs) when babies heard speech sounds. Two- to three-month-old infants heard a repetition of the same syllable: *ba-ba-ba-ba-ba*. At the first *ba*, the machine detected a wave of activity in the temporal lobe, the area where speech is processed. After the third *ba*, the intensity of this "detection wave" decreased, showing that the babies' attention was drifting off. They had *habituated to*—the term we use for "become accustomed to"—hearing the *ba*-syllable.

Then the investigators played the sequence of *ba-ba-ba-ba-ga*. When the babies heard the new syllable, the detection wave jumped up to the original intensity. The babies had heard something that was new to them. They could clearly distinguish the *ba* from the *ga*.

The investigators made a further discovery. Just after the EEG showed that a baby detected the difference in sound, a new wave of electrical activity appeared over his frontal lobe. Comparable observations of babies' reactions to other new sensory impressions show a similar pattern of electrical activity. First a detection wave appears in the specialized primary cortical area. This is followed by a "novelty" wave in the frontal lobe, indicating that the frontal lobe plays an important role in detecting novelty.

You may wonder if novelty will always win out. Not necessarily. In the little game you play with your four-month-old, his task is simply to distinguish between two relatively "neutral" items. He can keep a short-term memory trace long enough to detect which toy is new, and he wants to explore the new one. However, as he gains experience with objects, he will also use other characteristics as a basis for his decisions. An object's emotional value will become important. He may show more interest in the familiar and comforting ragged bunny that he takes to bed than in a new stuffed animal.

Forming categories

One of the exciting built-in features of your baby's brain is the ability to combine information into packages, or "categories," that he can compare as units, a process that aids learning. Even in their first few months, infants show that they are beginning to form categories based on physical appearance.

One group of investigators showed three- and four-month-old babies realistic pictures of cats and dogs. First they showed the babies seven pictures of cats one after the other. The cats were of different colors, and some were curled up, some were standing. After seeing a few of the pictures, the babies lost interest and didn't look very long when the next cat appeared. But when the babies saw a picture of a dog, they suddenly looked longer at the dog, showing that they perceived the dog as something new. Distinguishing between the cats and the dog meant that the babies had to group the common features in the pictures of the cats, most likely the more rounded faces, as one category, and to compare this to the longer face belonging to the dog. The dog's longer face shape did not fit into the "cat" category.

When seven-month-olds see a picture of birds with outspread wings and one of planes, they treat them as one category based on visual similarities (the wings). But 9- to 11-month-old babies treat them as two different categories. They have seen pictures of airplanes, seen them pointed out in the sky. They have heard birds singing and seen them pecking the ground for food. Infants begin to group items not only according to how they look, but also according to what they do. Comparing different categories according to both appearance and function is the beginning of forming "concepts" about the world.

Out of sight, still in mind?
Jean Piaget, the famous Swiss biologist and pioneer of child psychology, learned a lot about how children develop by playing with his own children. Once he showed his five-month-old daughter a small toy and let her play with it for a few seconds. Then he gently removed the toy and placed it out of her sight. She didn't look for it. To her, it was "gone." He tried this again a couple of months later, and she wasn't so easy to fool. She looked for the toy. Piaget said she had now recognized that objects continued to exist even when she didn't see them. He called this the concept of "object permanence."

Piaget was also the first to observe that infants younger than seven and a half months of age fail to retrieve a hidden object after a short delay period if the object's location is changed from the one where the baby has previously successfully located it. An important reason why a nine-month-old baby is better able to find a hidden toy is that he has developed a "magic" tool called *working memory*. This allows him to keep a picture, a "representation," of the toy's location online during the time when he doesn't see it.

Psychologists today believe that five-month-old infants have already developed a sense of object permanence and that other reasons can explain why they have difficulties in finding objects that are not in view. For example, the babies might not be able to use their arms for reaching, they may have more difficulty carrying out a plan of action, and they may not be as capable of keeping the object's location in mind as an older infant.

To follow the development of working memory in infants, psychologists use a little game called the A-not-B-task (introduced by Piaget) and it is fun for the babies to play. You can even do it at home every few weeks to see how your baby's memory for hidden objects improves. In the first step of the A-not-B-task, an examiner shows a baby two cups and hides an attractive toy in one of them (location A). Then he covers each cup with a cloth so the baby can't see the toy anymore. After one or more seconds he lets the baby reach to one of the cups to find the toy. The baby usually reaches toward the cup where he saw the examiner hide the toy. The examiner plays the game a few times with the baby, always putting the toy in the same location A. Then, right in front of the infant's watching eyes, he puts the toy into the second cup (location B) and covers the cups with the cloths. If the baby is allowed to reach immediately for the toy, he finds it.

But now the examiner makes the baby wait a few seconds and distracts him from looking at the cups, before he is allowed to reach for the toy. This time most six-month-old infants reach over to the wrong location A and, of course, don't find the toy there. This is called the A-not-B error.

In order to find the toy in the right location, the baby has to keep the memory of the toy's location online during the delay.

This is more difficult than in tests of recognition memory, where the baby sees two objects at the same time and is expected to detect which one is new. Now no cue is present. So he needs the help of working memory to keep the image (representation) of the toy and its location in mind.

As babies get older, they are able to remember where the toy is hidden in the A-not-B-task for longer periods of time. Adele Diamond observed that at seven months, babies were able to wait for 2 seconds, and at one year they were able to wait all of 10 seconds. Martha Ann Bell and Nathan Fox found that the increase in the length of time babies can remember is reflected in changes in electrical activity in their frontal lobes, showing that the frontal lobes play a role in working memory.

Results of research on brain function and working memory in infant monkeys support the findings in human children. When monkey babies played a game similar to the A-not-B-task, their memory span also increased with age and was accompanied by changes in activity in their frontal lobes. The scientists discovered that the main activity was specifically located in the front part of the brain just behind the forehead, the prefrontal cortex. A special group of neurons in the prefrontal cortex got very active during the time the monkey had to wait before reaching for his reward. The firing of these neurons could be what is keeping the memory online during the delay.

Adele Diamond brought up a very important aspect of the A-not B-task. She observed that many of the younger babies actually glance first to the correct location before they reach to the wrong place. They are apparently not able to resist their impulse to reach to the place where they had found the toy before. However, two months later, they are able to resist their automatic impulse to reach to the wrong location, showing that they are becoming able to guide their own behavior in order to reach a goal (the toy). This ability requires the cooperation of the prefrontal cortex, hippocampus, and motor cortex. The baby has to be able to form memories, make a "decision," and tell his muscles what to do.

During the second half of the first year, development in the baby's prefrontal cortex is in full swing. One of the ways we know this is by measuring how much energy it is consuming. New imaging methods make it possible to measure how much glucose, the brain's fuel, is taken up by a particular area of the brain. The prefrontal cortex consumes an increasing amount of glucose during the second half of the first year. The neurons there are making lots of new connections and growing more receptors for the neurotransmitter glutamate, a messenger substance necessary for learning. In addition, the transmission of electrical signals is becoming more efficient.

Memories in action

One way babies learn is by making an association between something they do and what happens as a result of their action. If they find the result pleasant, they will be more likely to repeat the action. Carolyn Rovee-Collier and her colleagues studied the development of a baby's ability to form associations, keep memories, and act according to them.

Babies often kick their legs in the air just for the fun of it. They smile gleefully when they see a mobile jiggle. So Rovee-Collier tied a ribbon around the foot of the baby and fastened it to a mobile hanging over the child's crib. When the baby kicked his leg fast, the mobile was set in motion, to the great delight of the baby. Babies quickly made the association between vigorous kicking of the leg and the movement of the mobile, and they kicked more often.

The older the babies were, the longer they could remember the mobile trick. At two months it was 1 day. At three months the babies could remember for 8 days. And the six-month-olds could even remember the trick for 14 days. What was going on in the babies' brains to make them better able to remember?

One answer might be that the younger babies were simply not as good at forming memories in the first place. This was not the case because all three age groups in the study learned the mobile trick initially in the same amount of time, so the younger infants were indeed as capable as the older ones of establishing the relationship between their kicking and the movements of the mobile.

Did the younger babies' memories just get deleted, wiped out like chalk on a blackboard? Or were the babies simply unable to retrieve those memories in order to use them? To find out, the investigators trained a new group of three-month-old babies to activate the mobile. Since they had observed that three-month-olds don't remember the trick after eight days, they let these babies wait a whole month so that they had ample time to forget their training.

Then, on the day before the test, half of the babies were placed in the mobile-furnished crib, and the investigator set the mobile in motion. The babies simply watched. On the following day, all the babies were returned to the crib with the mobile. Only the babies who had seen the mobile move the day before started their fast kicking as soon as they were placed in the crib. So the memory was there all the time. The baby just needed prompting to get it back. The presentation of a cue, or hint, is called *priming*, and it is something we can all observe in our everyday lives. Maybe you forgot your mother's birthday, but when you pass the flower shop, you suddenly remember.

Learning by observation

Actions of other people have a magical attraction for babies. Some scientists have suggested that even newborn babies try to imitate the mouth movements of adults. Imitating others is an efficient means of learning new behaviors and acquiring language. Between the ages of 9 and 12 months, babies begin to imitate actions they see performed in daily life. It's fun for them to take a brush and try to brush their hair, to put a telephone to their ear and start "chatting," or to grab a pencil and push it around on a piece of paper.

From around the middle of their first year, babies not only imitate actions immediately after they see other people perform them, they can even remember what they have seen and reproduce these actions later. Rachael Collie and Harlene Hayne showed six-month-old infants little toys on a felt-board that was just far enough away so the babies couldn't touch it. The experimenter then selected one of the items from the board and played with it in a particular way. For example, she pulled a cord to make a puppet's legs jump. Then she selected some different items and played with them. She repeated each activity six times so the baby had time to take note of the object and what she did with it.

Twenty-four hours later, both the experimental group and a group of babies who had not seen the toys before were allowed to play with the felt-board themselves. The babies who had seen the examiner play with the toys clearly played more with the items they had seen used the day before. And they correctly imitated the actions they had seen. The babies were able to learn and remember without actually handling the objects or going through the motions themselves.

Babies don't even have to see a real person perform the action in front of them. Andrew Meltzoff showed nine-month-old babies a television sequence in which a person manipulated two colored blocks. After watching him, the children imitated his actions. And they even repeated them 24 hours later.

Learning through play

For babies, exploring the world is all play, and their play changes as they grow. At first, the touch receptors in and around the mouth are important. Much of the baby's exploration is oral. Later it becomes more manual and visual. At around six months, infants manipulate just one object at a time, turning it over, looking at it intensively, and usually fitting it into their mouths. A few months later, they are already beginning to combine or relate two separate objects, blocks for instance, and bang them together.

Babies repeat actions discovered by chance when they have attractive consequences, like shaking a rattle or dropping keys to the stone floor to make a noise. This is the phase of dropping keys onto the floor, flipping spinach off the spoon, and using a toy hammer to pound in large, round pegs. Exhausted parents have to keep reminding themselves that babies are learning valuable lessons in cause and effect.

An interesting experiment showed that toward the end of their first year, infants are able to carry out a sequence of steps to reach a goal, a "means–end" procedure. The babies saw a toy resting on the table well out of their reach. Just within their reach was a cloth, whose most distant corner had a string attached to the toy. A foam barrier was put up between the baby and the end of the cloth nearest to him. Twelve-month-old babies were able to remove the barrier, pull the cloth, retrieve the string, and pull it to obtain the toy.

Around this time a baby begins to show first signs of a "mastery" smile, an expression of pleasure at being able to accomplish a task for the first time. His smile seems to say, "I did it!" The mastery smile represents a further step on the way to self-awareness. Once a task becomes a sure thing and loses its status as a challenge, it ceases to merit special attention, and the baby no longer smiles when he is able to do it.

The discoveries your baby is making are truly amazing. He has learned that objects have specific physical properties and something about how they are located in space. He has found out that his action can result in a pleasing effect. His working memory helps to store his impressions and compare them with new ones. The amount of information he can retain expands. He is beginning to resist his automatic impulses and to put simple plans and strategies to work. Around every corner he finds new and fascinating objects to investigate. He has a great time making sense of his world.

The Appearance of Fear

Easy-going Emily was a cheerful, sociable baby who delighted everyone around her with her sparkly eyes and sunny smiles. Even strangers could pick her up. So it was a shock to Deborah when, at the age of about seven months, Emily put up a huge fuss when Allen's mother came for a visit. The adoring grandmother had just returned from a winter in Florida and wanted nothing more than to hold her beloved Emily in her arms again. As she approached her little sunshine, Emily observed her cautiously for a second, and then a frown spread like a dark cloud over her face. A few short gasps, and out it came—a sorrowful wail that tore at Grandmother's heart. Deborah had to console two upset people at once!

At around 7 to 10 months, most babies begin to react differently, at times quite negatively, to the attentions of strangers or to being separated from their caregivers, in most cases their mother. This behavior is independent of cultural setting, as infants growing up in Africa, Central America, Europe, the United States, or the Middle East all react in a similar manner. It doesn't seem to make a difference whether a child is carried around all day by the mother, left to the care of another adult, or spends most of his time in a day care center. The more unfamiliar the stranger or the situation, the more likely the infants are to show signs of fear.

By the time they are three to four months old, babies are perfectly capable of distinguishing between familiar and unfamiliar persons. However, three- to six-month-old babies smile in delight at almost everybody. Why, at the age of around nine months, does the baby find the approach of a stranger suddenly so upsetting? Since the behavior is universal, we can assume that it has something to do with changes that are going on in the baby's brain. One possibility is that it involves the baby's improved capability of "working memory."

When the older baby sees a stranger approach, he is now more able to rapidly compare the stranger's features with stored memories of familiar persons. In order to do this, he must retrieve the representation—of his mother, for example—from his memory store and keep it online while he compares it to the image of the stranger. The comparison is unsettling. So the baby is afraid and begins to cry.

The baby's distress at being separated from his mother

could be due to a similar process. When a mother leaves her child, the baby compares the familiar situation where his mother was present to the unfamiliar situation when he is suddenly left alone. This creates uncertainty and, like the presence of a stranger, can lead to fear. The baby's face takes on a gloomy expression, and he cries.

In addition to the development of working memory, another important change is taking place in a baby's nervous system. During the second half of the baby's first year, stronger ties are being formed between the baby's cortex and amygdala, the structure that goes into action in time of danger. The cortex can now contribute more of its information. Right now this message is, "Watch out, I haven't seen this person before," or "My mother is going to leave me." The connections are also strengthened in the opposite direction. The amygdala activates the baby's cortex, making the whole system more sensitive.

The involvement of the cortex has been shown in some fascinating studies of children's individual reactions upon being separated from their caregivers. When mothers leave their 10-month-old children and go out of the room, some babies begin to cry almost instantly. Others hesitate before they signal their displeasure. These groups are comparable to the high and low reactive children seen by the Harvard Infant Study at the age of four months. It is most interesting that the two groups show differences in their patterns of electric activity in the brain even *before* their caregivers leave the room. EEG investigations by Richard J. Davidson, Nathan Fox, Martha A. Bell, and Nancy Jones showed that babies with greater activity in the right frontal lobe were more apt to cry instantly, while those with greater activity in the left frontal lobe tended to wait longer before crying.

The investigators also observed the different EEG responses depending on whether 10-month-olds were approached by a stranger or by the mother. When a stranger appeared, the infants were more likely to exhibit greater relative activity in the right frontal lobe. When their mother appeared, the greater response was on the left.

Whether or not a baby shows fear does not necessarily depend on the event itself but often on whether or not he has any control of the situation. If a stranger rushes up and grabs the baby before he has had time to size him up, we can hardly blame the infant for putting up a fuss. Studies have shown that babies are less afraid of going into a strange room or of approaching strangers when they do so on their own initiative. One study showed that a toy that usually frightened one-year-old infants could be transformed into a pleasant toy if the baby could control the toy's actions. Half of the children had a chance to set a cymbal-clapping toy monkey in motion by hitting a panel taped to a tray. For the others, the monkey began clapping by itself. The children who could control the monkey themselves were less frightened and even smiled at the monkey's antics.

Challenge and distress

From the very beginning, a baby's bright new world has occasional clouds. The first can be caused by hunger pangs, the unavoidable momentary pain of an inoculation, the discomfort of a wet diaper. As your baby grows, so does the list of potentially disturbing events. It grows to include not only immediate physical sensations but also uncertainty about what might happen, for example, when Mother leaves the room or a stranger appears.

Part of a baby's experience in his new life is learning to deal with these challenges. As we have seen in Chapter 3, babies begin with individual levels of sensitivity to unfamiliar events. However, their biological and cognitive development, together with the nature of their experiences in daily life, influence their manner of responding to them.

Megan Gunnar and her colleagues in Minnesota made use of routine medical procedures to study changes in the way babies react to upsetting events, ones that could be expected to represent a mild stress for the baby. When the babies came for their periodic well-baby checkups, they were undressed and put on the scales to check their weight, a harmless enough procedure, but one that babies don't relish. The investigators noted the babies' behavior and measured the cortisol level in the babies' saliva at this time. When a baby is faced with an upsetting situation, his hypothalamus orders his adrenal glands to step up the output of the hormone cortisol, making this a measure of how his body registers stress.

From birth to three months, when they came for their checkups, the babies' salivary cortisol rose intermittently to high levels. However, at around three months of age, the baby's body reacted less intensely, and the cortisol responses became more stable. This—like the reduction of crying and fussing at about this time—could be due to the fact that the cerebral cortex is beginning to take control over subcortical structures. From age 3 to 15 months, the cortisol responses to mild stress gradually decreased.

When the babies were six months old, the investigators made an interesting observation. The babies' cortisol response to the undressing and weighing was gradually going down, so the researchers could assume that the babies' bodies were feeling the stress less acutely than before. Yet in spite of this, the infants began to protest more vigorously and vociferously than ever. This may be because new cognitive abilities are entering the picture. A baby is more likely to experience a sense of "déjà vu." He connects the feeling of the room with a series of unpleasant events: getting undressed, being weighed, getting an inoculation, being approached by a stranger. He is also beginning to learn that his reactions can have an influence on other persons. So he protests loudly.

To Think About

Why is a pediatric checkup important?

Because this is the time when so much is going on, babies should have regular checkups. A comprehensive medical checkup comprises general health, physical growth, sensory and motor development, vocalizations, sleep and feeding habits, play, and social behavior. It is a wonderful occasion for you to ask the baby's pediatrician about anything that interests or bothers you.

What can I watch for while playing with my baby?

Playing together is a wonderful time to get to know your baby and observe what he can do. Inevitably you will compare your baby to others around you, or you will read in the newspaper or hear from your mother what babies are supposed to do at a certain age. In our hectic world with its demands to be ever better, ever faster, more efficient, it is extremely important to remember that babies need time to develop according to their own individual pace of brain development. They need time to absorb all their new impressions and make discoveries on their own. Don't rush them. However, if anything troubles you, don't hesitate to have a talk with the baby's pediatrician.

Watch how your baby discovers his own body and learns to coordinate his muscles and senses. Simple motions reveal that your baby is exercising very important functions. Touching one hand to another is the beginning of learning to use both hands at the same time. When he looks at his hands, he is practicing the coordination of eye and hand.

Observe which kind of plaything your baby likes most. These are not necessarily the "educational" ones bought in the store. Some babies prefer soft, cuddly animals, while others are more interested in your key chain or other household utensils. Infants are learning to combine information from all their senses. They enjoy hearing new sounds, feeling textures and shapes, and looking at bright colors. Babies show more interest in things that are new, so it is better to let them explore a few toys at a time rather than surrounding them with all their toys at once.

Since one-year-olds are becoming more adept at figuring out how to retrieve items that interest them, you can devise little "games" to help your child develop strategies for reaching his goal. Try putting an obstacle in front of a toy for the baby to remove, or put the toy in a box. By presenting tasks that require an increasing amount of effort, you can encourage your baby's focused attention and persistence. Keep your sense of humor. Your baby may bring in his own ideas of the game, as when he pulls a tablecloth to get a glass of milk or empties your pocketbook to see what's inside.

Does music have an effect on my baby's brain development?

What are we to make of the profusion of tapes and CDs of classical music claiming to "build your baby's brain"? No one would dispute that gentle strains bring pleasure and relaxation to both infants and adults. Some adults experience a state of relaxed attentiveness while listening to music and feel more productive or creative. However, there is no credible evidence that listening to music makes the brain grow faster.

If you ever dreamed of success as a singer, this is your chance. For your baby, whether you are a shower soprano or a bathtub baritone, you are number 1 on the Hit Parade. It doesn't matter whether you remember the words—in fact, you can make them up as you go along. Singing is communication. It's relaxing. And Baby loves it.

How can I make my baby's surroundings safe to explore?

A baby needs safe spaces to crawl around in, where he can't fall down stairs, get his hands on items small enough to choke on, or come in contact with electric currents. Being able to explore a room gives him an idea of space and lets him exercise his muscles. But he doesn't have to roam freely all the time. A short time in a playpen, where he can see and hear what people are doing around him, is an opportunity to concentrate with fewer distractions on some interesting toys, while you concentrate on answering the mail or cooking dinner.

By the end of the first year, babies can play well using both hands and are practicing picking up small items with thumb and forefinger. Since a lot of exploration takes place by mouth, dangerous items should be kept well out of reach. This means anything that is breakable, fits entirely into the baby's mouth, has sharp edges, or is painted with a toxic substance. Watch out for pins, thumbtacks, paper clips—and peanuts.

Should I start teaching my baby to read?

Not long ago, a program on the BBC's Learning Channel showed a mother presenting her six-month-old daughter with large white cards with simple words like *dog* and *cat* written on them in big, bold letters and speaking the words slowly and clearly to the infant. A short time later, I found a book called *How to Teach Your Baby to Read*, whose authors suggest that parents should begin to teach their babies to read as soon as their eyes can focus on the large letters. The book even claimed that this practice would stimulate the baby's sense of vision. While a baby might be mildly amused by his mother's attentions, no evidence indicates that babies need visual training or that teaching efforts of this sort will lead to good reading habits later.

Comfort and Communication

Babies are born into a world of other people, ready to form social ties, first to their primary caregivers and later to others in their surroundings. Close, caring relationships give them the security they need to find their way in their new world. It makes sense to refer to the ties an infant forms with her caregivers as attachment, yet when we try to pinpoint exactly what this phenomenon is, it eludes our grasp. We simply have no means sufficient to measure it.

Primates, including human beings, nurture their infants in close body contact. Mothers not only nurse their babies at the breast but also fondle and caress their infants. Holding and touching the baby foster powerful emotional bonds. This means that babies will be most likely to look to their caregivers in times of uncertainty and distress. These are the people best able to calm their fears and comfort them. In this sense, human babies become attached to their fathers as well because fathers also hold them and cuddle them.

Children need a warm, consistent environment in order to form close relationships with people around them. These relationships are not established in a few weeks but, rather, take years to grow. Strong bonds can also form between parents and their adopted children.

If caregivers ignore children or do not respond appropriately to their needs, these children may have learning difficulties and trouble forming relationships with other people later. Studies have shown that a baby's interactions with caregivers can have an effect on the infant's stress level. In one investigation, nine-month-olds were separated from their mothers for 30 minutes and left with an unfamiliar baby-sitter. If the baby-sitter was friendly, playful, and sensitive, the babies did not show any increase in salivary cortisol. However, if she remained cold and distant, their cortisol levels went up, particularly those of children described as more negative in mood or more easily upset. These children stood to benefit the most from sensitive, responsive substitute care.

How a child responds to a particular stressful event is subject to many factors. Differences in temperament are one reason one child reacts more intensely than another. However, as a child grows, her previous experience takes on ever-greater importance. A healthy relationship between the child and her caregivers provides a secure basis from which she can learn to deal with upsetting events and recover her own sense of balance.

Song without Words

Babies don't only cry. They are born with a variety of "languages" that enable them to engage in social interactions long before they utter their first real words. The first interactions with the world are emotional in nature. Infants respond to touch, to a firm, warm hug. They are comforted by soft words or a gentle lullaby. They express their discomfort or distress by waving their arms and legs. As their systems become tuned to perceiving and integrating visual information, they become able to read the emotional signals in facial expressions. By the end of their first year, just as they are beginning to take their first real steps alone, they are also ready to set out into the world of words.

If you watch a mother speaking with her two-month-old baby, you will not be surprised that psychologists have employed musical terminology to refer to this "duet" of mother and child. The mother speaks softly, and the baby gazes up at her in rapt attention, as if listening carefully to every word. Then the mother pauses, waiting for an answering smile, a cooing sound. The psychologist Daniel Stern calls this "attunement." It is a dialogue without words, a sharing of an emotional experience.

Even without words, babies establish their lines of communication by watching and echoing facial expressions of emotion. Edward Tronick and Jeffrey Cohn observed how often mothers and their babies used the same expressions at the same time (matching) and how often they changed their behavior with respect to one another. The youngest babies were three months old, the oldest nine months. Each mother was seated facing her baby and put on either a happy or a sad face. When the mother smiled, her baby was more interested and also showed a happy expression. And when she frowned, the baby adopted a more negative expression. The matching was a kind of dialogue.

Surprisingly, the authors of the study found that two-thirds of the time the expressions of the babies and mothers didn't match at first. However, within a few minutes, babies and mothers coordinated their expressions. Infants as well as mothers initiated the adjustments. The ability to coordinate emotional expressions increased with the baby's age. Being able to sense and react to another person's expression of emotion is a crucial early step in communicating with other people. Babies learn about others and their emotional expressions when they have the opportunity to watch, listen, and participate in social interactions.

That special smile

Since smiling is so important in human life, it is surprising it doesn't get more attention than it does. Distress and fear are far more likely to be research subjects than smiling or laughter. If we consider a baby's social smile as a type of care-eliciting behavior, why isn't it present at birth? Its absence may have something to do with the interesting speculation that human beings are born a few months prematurely, from an evolutionary point of view.

At birth, a baby's facial muscles are already prepared to make a smile. We see this because every now and then the baby's lips stretch out in a fleeting expression reminiscent of a grin. This is a reflex, however, not an expression of pleasure.

When your baby is around three weeks old, you can tease the first glimmer of a smile out of her by gently brushing your finger over her cheek. At first, only the corners of her mouth turn up ever so slightly. A week later, she may show a tentative smile when you nod your head in front of her.

Any minute now it happens—that special, unmistakable, first real smile. Your baby will look up at you, and this time not only her lips, but also her eyes get into the act. Now she is putting her whole show together, so to speak. Her senses of vision and hearing are more acute, and she is becoming an active participant in the "dialogue." Her whole face lights up in delight.

Peekaboo!

The simple game of peekaboo has universal appeal for babies and parents alike. It usually goes like this: When your baby is about six months old and happily seated in her infant seat, you slowly lift your hands to cover your face. A wisp of a cloud flits over the baby's features—a split second of uncertainty—and then she grins from ear to ear as you flip your hands back to reveal your smiling face. It won't be long before she tries to initiate the game herself by lifting her hands toward her face in an irresistible invitation to participate.

In their review of smiling in infancy, L. Alan Sroufe and Everett Waters at the University of Minnesota, suggested that there is more to smiling than a sign of social pleasure—as if this were not enough. Smiling is also related to an infant's ability to cope with novelty. A baby is able to detect that a situation is new.

She senses a momentary uncertainty about it, followed by a release of tension when she is able to fit it into her scheme of things. A similar mechanism may underlie her "mastery smile," when she realizes that she can accomplish something on her own, such as standing up unsupported or being able to reach a favorite toy.

Babies not only smile. They can also laugh. Around the age of four months, the surest way to get a laugh is to make funny little *brrr*-sounds and kiss or tickle your baby's stomach. But about two months later, she will begin to laugh even *before* you start to tickle her stomach. She shows pleasure in anticipation of the fun.

Toward the end of the first year babies begin to laugh at funny situations; for example, when a mother waddles across the room pretending to be a penguin. Recognizing a funny situation is quite an achievement for the baby's brain. The baby notices that the situation is not only new, but there is also something odd about it. She is puzzled for a moment because it doesn't correspond to what she thinks it should be. However, she soon realizes that it really is her mother, and she laughs at the joke. It's a bit like peekaboo, when Mother's face disappeared for a moment and then miraculously reappeared. Her momentary uncertainty is relieved, and reassurance takes its place.

Reading the signals

Sometimes you wonder why your baby picks the least appropriate moment for a fussy spell. It often seems as if babies have a sixth sense for picking up the tension in the air. And you are right. They sense the rise in your voice, the quickening of your step, the tension in your touch. The fact that babies are sensitive to the behavior of adults around them can also have a positive effect, for example, on how infants cope with novelty. Studies have shown that the way a parent approaches a stranger has an effect on the child's expression of fear of strangers, a behavior that typically emerges in most infants around the age of 7 to 10 months. Babies, who are great imitators, respond less negatively or even positively to strangers when their accompanying parent takes a positive approach. Greet the new baby-sitter with a friendly smile and speak with her in a relaxed tone of voice, without pushing your baby into close contact with her too early.

Toward the end of their first year, babies not only "sense" the reactions of their parents; they also actively refer to them as if using signals from their parents as a beacon to guide their behavior. If your baby sees a camera lying on the coffee table and you tell her not to touch it, she may hesitate a moment and look at you before reaching out her hand. She looks as though she is saying, "I'm not quite sure about this. What do you think?"

Infants can be very individual in how often they glance at their parents in times of uncertainty and in how they take the "hint," depending on their temperament, age, gender, ability to focus attention, and past experiences in similar situations. Studies have shown that babies refer equally to mothers and fathers. "Social referencing" is an important feature in the hundreds of adult–child interactions that take place every day. Through them, children model their behavior and learn through experience to distinguish between actions that get positive feedback and those that are discouraged. Social referencing is a basis for the formation of standards.

From Sounds to Words

When we speak to newborns, we act as if they could understand every word we say. We repeat important words and emphasize them, accompanying our words with particularly broad smiles of delight and encouragement: "You are such a good baby!" Of course, the baby doesn't know what the single words mean. But as she grows, she quickly gets an idea of the feelings behind the words, much as we can sense the feeling behind a singer's rendition of an operatic aria in a foreign language. The melody of speech conveys an emotional message. A rising intonation gets more attention, as in "Wake up, it's time to get up!" A singsong melody with a higher pitch and soft, swelling volume is encouraging: "Oh, that's a *very* good ba-by!" A gentle, falling cadence is comforting:

"Now it's time to go to sleep." And a sequence of short words with sharp consonants shows disapproval: "Not again! Stop it!"

Babies pick up more than just the feelings behind speech. They are also acquiring basic capabilities that set the stage for understanding and using the building blocks of language. The special kind of language people direct at infants (now called *parentese* to distinguish it from frivolous baby talk) may help them learn about how the basic units of their native language are put together. With its exaggerated emphasis of important initial sounds and clear intonation of phrases, with longer pauses at the end of sentences, parentese gives babies stronger language cues than those available in adult-directed speech.

At birth an infant's brain already registers certain differences in speech sounds, an ability that will be important later for the baby's understanding of spoken words. She will have to recognize the sounds that are crucial for the meaning of the word. The smallest units of a language that can change the meaning of a word are called *phonemes*. Some examples in the English language are the *b* of *big* and the *p* of *pig*.

Until about the age of six months, babies are also able to detect differences in sounds that occur in other languages. Experience then shapes the range of sounds they can discriminate, and by 12 months they are more sensitive to the sounds of their native language than to foreign sounds. Four-month-old Japanese babies, for example, can detect the difference between *l* and *r*, but they gradually lose this ability, since they do not hear the *r* spoken in their environment. However, this doesn't mean they can't learn, with proper training, to discriminate the sounds later.

Cooing and babbling

From the moment of birth, infants make it abundantly clear to us that they can indeed produce sound. However, they need time to explore the potential of this powerful instrument. At first parents may be able to distinguish the sharp burst of a cry of pain from the more drawn out cries that signal hunger. As time goes by, a baby's cry may become a more specific call for personal attention. She feels lonely, bored, or wishes to have a toy to play with. But it would be unfair to suggest that babies only cry. Even newborns make little fussing noises to signal mild disturbances, or sighs and soft murmuring sounds of contentment.

At around the age of three to four months, a baby begins to form soft, cooing sounds, little "oohs" and "aahs," often preceded by a consonant. By this time her larynx has descended to its adult position, and she is now better able to control her breathing and the muscles of her tongue and mouth. Babies often coo when no one else is around. This may be their way of playing with sounds, and babies, who are not alone in this respect, seem to enjoy hearing the sound of their own voices.

With cooing, babies gain a new means of communication. When you coo at your baby, she purses or stretches her lips in an attempt to imitate the sound. Then she begins to start up the "conversation" herself by cooing at you first. The two of you are soon taking turns "speaking" and listening to each other.

Cooing is a kind of musical speech. Around the time babies begin to coo, major development is taking place in the area of the baby's right brain hemisphere corresponding to the language area in the left hemisphere. The neurons rapidly sprout a profusion of dendrites, which grow longer and make contact with other neurons. The right hemisphere is known to be more involved in processing the melodic and emotional qualities of words, while the left contains the lexical items (vocabulary) and the center that programs articulation. In most people, language functions are located mainly in the left hemisphere, a situation that is related to handedness. Language functions are concentrated in the left hemisphere in 96 percent of the right-handers. While 70 percent of the left-handers also have left hemispheric speech centers, the remaining 30 percent have language centers that are either located on the right or distributed over both hemispheres.

Richard J. Davidson and Kenneth Hugdahl have suggested that mental functions aimed at specific points in external space—such as looking and listening—are represented symmetrically in both hemispheres of the brain. Operations not related to a specific location in space—such as language, emotions, problem solving—are not represented symmetrically. Instead, they have highly specialized components on one side or the other. Being able to mobilize these specialized units may allow the brain more ways to process information.

Babies' first language endeavors are similar, whether they are born in the United States, Switzerland, Brazil, or Indonesia. For this reason some scientists speculate that the first sounds babies make are those our ancestors used when they discovered they could communicate by sounds made by the mouth. Babies produce their first sounds by simple jaw movements, by opening and closing their mouth and moving their tongue. First come easy sounds like *ma-ma* and *ba-ba*. Then follow sounds that need the tongue, such as *da-da*. Sounds like *go-go* and *ko-ko* that need the muscles further back in the mouth come later.

Parents in Europe and the United States speak to their babies as soon as they are born—if not before—as if babies could understand every word they say. They don't, of course. However, they do begin to pick out words in a remarkably short time. A recent study at Johns Hopkins brought what seemed to be news to warm the heart of every new parent. The first words that babies recognize as belonging to specific persons or objects are "mommy" and "daddy." Six-month-old infants were seated on a parent's lap facing two television monitors. On one screen the baby saw a videotape of her mother and on the other a tape of her father. When a synthesized, gender-neutral voice called out either "mommy" or "daddy," the infants looked longer at the parent being named. It is only about two months later that babies identify words with objects. In a further two months, they will speak the words *ma-ma* and *da-da* all by themselves and mean them.

From babbling to first words

Why is it that babies understand many more words before they say them? This is because actually producing the sounds, even in imitation, is a very complex procedure. Babies not only have to recognize the sounds of a word; they have to put them together.

When infants hear a word repeatedly, they group the individual sounds of the words together. This may correspond to a linking of neurons forming a kind of "map" in the brain. With practice, the maps are adjusted and strengthened, making the baby better able to imitate sounds.

Between four and seven months, important changes are taking place in the brain that are crucial for the baby's improving language skills. The connections between the baby's auditory cortex, the region that processes what the baby hears, and the mouth region of her primary motor cortex are becoming more efficient. These new connections make her better able to listen to the sounds of speech and pattern her own sounds on what she hears.

Between 6 and 10 months, babies all over the world begin to add distinct syllables to their repertoire of cooing sounds and string them together in melodic sequences reminiscent of speech, called babbling. The onset of babbling does not seem to be directly linked to what an infant hears because deaf children also begin to babble normally around the age of six months. However, deaf children stop babbling when they are around nine months old, indicating that hearing is necessary for the further development of babbling into real words. The sounds of babbling are similar enough to the sounds used by adults in regular speech to make it seem as if the baby is carrying on a real conversation. While babbling starts off all over the world with sounds like *ma-ma* and *ba-ba*, babies soon adapt their babbling to the sounds and melodic features of their native language.

Producing the sounds of an infant's native language requires the precise control of her mouth and throat muscles, and both the muscles and their control centers in the brain need time to develop. The coordinated movements necessary for speech production have to be programmed in advance, the job of a special area in the baby's left hemisphere called Broca's area. This area is named after Pierre Paul Broca (1824–1880), the anthropologist and surgeon who discovered the connection between language deficits and abnormalities of this particular region of the cortex. As the language "boss," Broca's area tells the motor cortex which muscles to activate in order to produce the desired sounds, for example, *ba-ba-gi-go*. Around the age when babies start to babble, the neurons in Broca's area are rapidly developing. It's now the turn of the dendrites in this area to suddenly grow longer, branch out, and form new connections.

As babies approach the end of their first year, they actively put their new communication skills to use. They combine babbling with looking in a particular direction, and then suddenly, when you listen very hard, there it is, the very first word. Some babies come out with the magic words *mama* and *dada* before they are a year old; others keep their parents waiting for several months longer.

Babies learn to speak without any special stimulation, provided that they can hear language being spoken around them. However, the nature of the social environment around them does seem to play a role, though we may smile at the words of a writer on education cited in a book on child care written in 1831: "The heaviness of the Dutch and the vivacity of the French are owing to the different manner in which infants are treated in those two countries. The Dutch keep their children in a state of repose . . . ; the French are perpetually tossing them about, and showing them lively tricks."

A recent study by Barry S. Hewlett and Michael E. Lamb showed the effects of different parenting styles in two neighboring tribes in central Africa. The investigators found that Aka mothers hold their babies more often than the Ngandu mothers do, while the Ngandu mothers leave their babies alone for longer periods of time. The Aka babies generally cry and fuss less. But the Ngandu mothers are more likely to stimulate their babies by speaking to them, and their infants smile and vocalize more than the Aka infants.

In the short span of one year, babies make tremendous progress on their way to both autonomy and integration into the world of other people. Without being conscious of it, they are learning how they can affect the behavior of others. Is it the friendly smile or the loud cry that gets more attention? They are learning to interpret the emotional signals of others and using them to guide their behavior. Mother's grim expression means "no," her smile means "go ahead." Infants have made their first acquaintance with the basic tools of communication: gesture, facial expression, and the sounds of speech.

To Think About

Should I teach my baby sign language?

Recent articles in popular magazines have suggested that teaching infants sign language before they are able to speak would improve their ability to communicate, making them less frustrated and irritable. There is no proof of this hypothesis. Instead, it might be better for adults to spend their time paying attention to the wealth of capabilities babies already have. Babies learn to use facial expression by watching adults. They modulate their tone of voice and use gestures, such as turning their head away, waving off an undesired object, pointing at an attractive toy. Because babies learn by imitating, adults should give clear signals that a child can learn to interpret. Be consistent when you use yes and no and the corresponding facial expressions.

What can I do to calm my baby's fears?

Even if you don't always need to "whistle a happy tune," you can try to project a calm, confident attitude in situations in which your child might be afraid. Speak in a low tone and try not to confront a child with new objects or strangers too abruptly. If a child is afraid of a toy clown, you might leave the toy in a place where she can reach for it herself. Because of the makeup of their nervous system, some babies have a tendency to be more fearful than others. For this reason, they may need more time to become acquainted with a new baby-sitter.

You needn't be worried about the brief stress a child experiences during routine health procedures. Occasional short-term stressful events of daily life have no negative effects on the child's health. Comforting words and a hug are the best "medicine."

Should I leave a night-light on?

You may have heard of children who begin to cry when parents say good-night, turn off the light, and leave the room. This should come as no surprise. The baby is now capable of learning that when the light goes out, she will be left alone, and she is understandably unhappy about it. Before this pattern is established, you might try a special nighttime ritual. Turn off the light first, then sing or play quietly for a while together in the dark before you quietly leave the room. Turning off the light doesn't necessarily come to mean the company is going to leave. The baby learns that darkness can be fun too. This procedure is often quite effective.

Should I worry about day care?

Between 1975 and 1997 the number of mothers of children under six years of age working at least part-time outside the home increased from 30 percent to 62 percent. Most mothers return to work by the time their babies are three to five months old. This means that most children today spend their first seven years in a variety of child care situations. Therefore, the National Institute of Child Health and Human Development (NICHD) is currently conducting a long-term, comprehensive study of the effects of child care outside the home. The NICHD started with a group of 1,300 children, representative of the U.S. population, and followed most of them through the first seven years of life. They have now published preliminary results.

Not surprisingly, higher-quality day care was related to fewer reports of children's problem behaviors, higher cognitive performance, better language ability, and higher level of school readiness. Lower-quality care predicted more problem behaviors, lower cognitive and language ability, and lower school readiness scores.

Children's attachment to their parents was the same, whether the children stayed at home or spent many hours in day care.

One finding was especially significant. The investigators found that, in general, "family characteristics and the quality of the mother's relationship with the child were stronger predictors of children's development than child care factors." This was true whether the children spent a great deal of time in child care outside the home or were cared for primarily by their mothers. Family life is thus an extremely important factor in the child's development.

Maps and Milestones

Brain Maps:
Finding Your Way around the
Nervous System

F inding your way around the nervous system is a bit like finding your way around in a new city. It helps to start out with
the main features of the map and then proceed to the labyrinth of smaller streets and alleys before you come to the individual buildings. If we look at a lengthwise section of the human
body, we see that the nervous system has two main components,
the central nervous system and the peripheral nervous system.
The central nervous system is the "hub," composed of the brain,
brainstem, and spinal cord. All nerve connections outside of these
structures belong to the peripheral nervous system. We will be
looking primarily at the central nervous system. However, we
mustn't lose track of the fact that the whole nervous system is
wired together.

A main feature of the brain is its lengthwise division into two
hemispheres, a phenomenon that can already be detected at
around four months after fertilization. Although the two hemispheres share many functions, they are not a mirror image of each
other. They work together, with each hemisphere contributing its
special area of expertise.

THE CEREBRAL CORTEX

The outer covering of the brain is called the *cortex*. It is only a few millimeters thick, but it has a large surface area that is wrinkled up to fit inside the skull. This part of the brain has developed particularly in human beings during the course of evolution.

The cortex is divided into specialized areas, although these regions do not have any fixed borders or "walls" around them. An area at the back is specialized for vision, one on the side for hearing; a band over the middle is the *motor cortex*, which is responsible for voluntary movements of the muscles. Adjacent to the motor cortex is an area responsible for body sensations. All of these specialized areas are subdivided into primary and higher order areas. The primary area receives the signals and transmits them to the higher order areas, which interpret them and send them on to the motor areas, which give the commands for action. Some functions show a crossover of control pathways. For example, the right motor cortex directs the movements of the left side of the body, while the left motor cortex sends its commands to the right side. A similar crossover takes place in the sensory pathways.

Special areas of the cortex are the experts for handling language and speech. In most human beings these come to be clearly located in the left hemisphere. *Wernicke's area* is the specialist for understanding language. *Broca's area* is involved in language comprehension and is responsible for setting up the orders to the muscles to form the sounds of words.

One of the jobs of the *prefrontal cortex* is to compare stored memories with present situations. Another job is making plans and evaluating strategies for carrying them out (executive functions). A special part of the frontal cortex is the *orbitofrontal cortex*, which is closely linked to the brain's emotional centers and plays a role in our emotional and social perception of events. This connection is an important basis for establishing a balance among our feelings, our thoughts about them, and our actions.

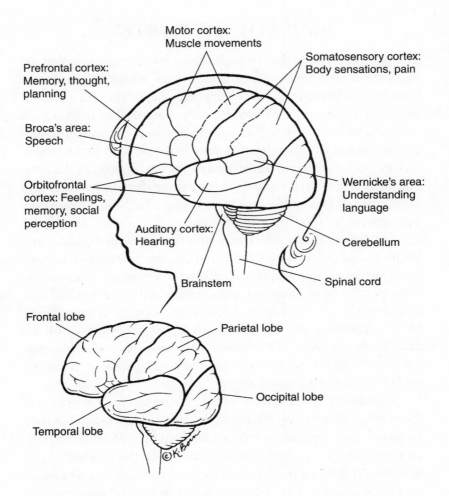

Similar to the way we use terms like port and starboard to describe positions on a boat, we use special words to localize structures in the brain. *Frontal* refers to the area in the forehead, *occipital* to the back of the head, *temporal* to the area in back of the temples, and *parietal* to the region between the frontal, occipital, and temporal areas.

BENEATH THE CORTEX

The brain structures that lie underneath the cortex are called the *subcortical structures*. The *hippocampus*, which gets its name from its sea-horse shape, is a temporary depository for information on its way to long-term memory storage. The information from the primary and higher sensory cortices travels to the hippocampus and closely related areas, where it is kept for up to several weeks before being transferred to specific regions in the cerebral cortex. The hippocampus may be more than just a temporary warehouse. It is possible that it serves as a facilitator for storing information in other areas in the brain.

The *basal ganglia* selectively turn on and off the specific motor programs necessary for the automatic performance of learned movements and for adapting to changes in posture. Because the basal ganglia have extensive connections with the cerebral cortex, they also play a role in higher order aspects of movement, such as planning the complex movements necessary to play a melody on the piano. The connections with the association cortex and the limbic system involve the basal ganglia in many aspects of both thinking and feeling.

Research on the functions of the *cerebellum* has long stood in the shadow of news highlighting the role of the cerebral cortex. However, this apparent neglect is about to be corrected. The cerebellum has strong connections to and from the cerebral hemispheres. Projections from associative and paralimbic cortices appear to be matched by cerebellar projections to the same regions, indicating that the cerebellum may be involved not only in motor functions, but also in cognitive abilities, language, and emotional responses. Clinical observations also point in this direction.

The *thalamus* is the main relay station between the cortex and the subcortical centers. It also functions as the entrance to the brain for information from the sensory organs. The *brainstem* is crucial for vital functions such as breathing and circulation. One section of the brainstem, the *reticular formation*, is particularly involved in alertness and attention. The *hypothalamus* plays a role in sleep–wake cycles. The hypothalamus and the *pituitary*

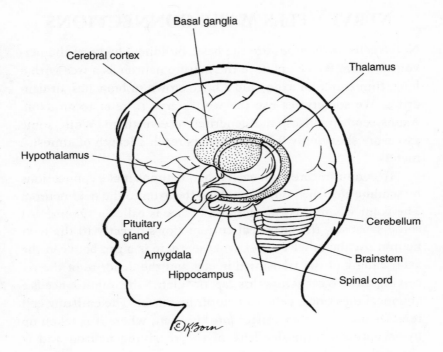

Basal ganglia

Cerebral cortex

Thalamus

Hypothalamus

Pituitary gland

Amygdala

Hippocampus

Cerebellum

Brainstem

Spinal cord

©KBorn

gland regulate hormone production. Together with the auto-nomic nervous system, they regulate the functions of our inner organs. They are therefore important for confronting stress and infections.

The almond-shaped *amygdala* is part of an ancient system, the limbic system, a feature that appeared early in evolution. The limbic system is important for basic survival because it reacts immediately to signs of danger by setting the body in alarm mode. In addition to heightening attention, it stimulates processes that prepare the body for both fast reactions and endurance. In human beings, the amygdala also contributes emotional significance to events. During the course of evolution, the human cortex has de-veloped immensely and become more closely linked to the limbic system. This means that the cortex can contribute more of its information, helping the brain to analyze a situation and seek more flexible means of dealing with it.

NERVE CELLS MAKE CONNECTIONS

Nerve cells, or *neurons* are the basic building blocks of the nervous system. We can picture the mature neuron as a tree with a long, thin trunk, or axon, and a huge crown of branches, or dendrites. We sometimes use the word nerve to refer to an axon. Axons send impulses, and dendrites receive them. While some axons are short (only 0.1 mm), others reach a length of almost 2 meters.

To communicate with each other, neurons make connections by sending electrical impulses along the axon to the next neuron. The point where this transfer takes place is called a *synapse*. At some synapses, the electrical current flows directly to the next neuron. At the other type of synapse, there is a gap between the axon ending of the transmitting cell and the dendrite of the receiving cell. To get across this gap, or "cleft," the impulse needs a chemical messenger called a *neurotransmitter*. The emitting cell releases the neurotransmitter into the cleft, where it is taken up by receptors on the dendrite of the receiving neuron and is changed again to an electrical impulse.

The neurotransmitters affect the actions of the nerve cells in many different ways. Each type of transmitter has its own special docking site, or *receptor*, whereby one neurotransmitter can even have several subtypes of receptor. This great variety allows a wide spectrum of complex behaviors.

If you think your list of e-mail correspondents is long, just compare it to the contacts of one single neuron. In your own brain, one nerve cell can have as many as 20,000 connections with other neurons. This means trillions of synapses in the brain as a whole, making it the most complex structure known to us in the universe. Because each neuron can have so many connections to other neurons, information can be widely distributed within the vast network of the brain. If all the connections of one adult brain were lined up in one long chain, it could be wrapped several times around the equator. Although the brain map shows neurons contacting other neurons, nerve cells also have contacts with other targets, such as muscles or glands.

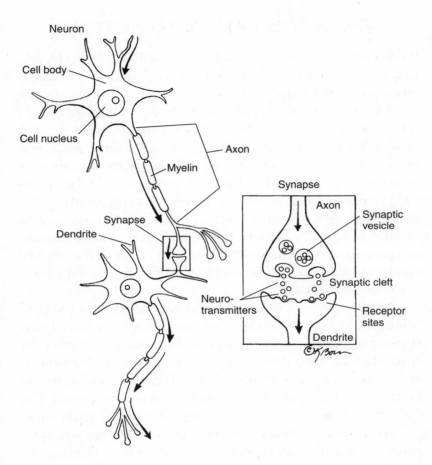

The axons of the neurons are surrounded by a myelin sheath, which insulates them, allowing electrical impulses to travel more swiftly and efficiently. Research published in 2000 has proven that myelination only takes place when electrical impulses travel over the axons. So brain activity stimulates myelination. It's a reciprocal process: more activity means more myelin, and more myelin increases activity. Although myelin is extremely important, some current gets through without myelin. Some connections—for example, axons with contacts to the internal organs—never get myelinated at all.

FROM PERCEPTION TO ACTION

We have chosen the example of a six-year-old boy and his bicycle to show the circuit of how the brain puts information together, forms plans, and gives the signals for action. The scene begins when the boy sees his bicycle leaning on the wall beside the garage door. His *visual cortex* analyzes the information projected from his retina over separate pathways for color, form, depth, and motion. The pathways for color perception and for the detection of shapes both terminate in the *temporal cortex*. The latter system is sensitive to the outlines of objects and to the details within them. Since this system is concerned with the features used to identify an object, it is called the "what" pathway. The separate components of the image—the color, the shape of the wheels, saddle, and handlebars—are brought together, and the boy's brain perceives his bicycle.

Another pathway in the visual system determines the location of the bicycle. The "where" pathway detects the spatial relationship of objects to each other and the motion of an object in space. This pathway projects to the *parietal cortex*. The parietal cortex interprets the signals it receives from the visual cortex to locate the bicycle in the driveway in front of the boy's house. The "where" and "what" pathways converge in the prefrontal cortex.

The *prefrontal cortex* integrates the information and compares it to situations in the past. Our little cyclist recognizes the bicycle as his own, the one he got for his birthday. It is bright red, with shiny black tires and a special bell. He is reminded of what fun it is to pedal down the driveway and feel the wheels turning smoothly over the asphalt. He decides to put on his helmet and take the bicycle by the handlebars, holding it in a steady vertical position so that he can keep his balance on the saddle. His prefrontal cortex sends notice of his intent to the *premotor cortex*, which prepares the orders for the necessary movements. From there, signals go to his *primary motor cortex*, which sends them to the neurons of his *spinal cord*. These neurons convey the orders to the muscles, putting them into action. He starts his rhyth-

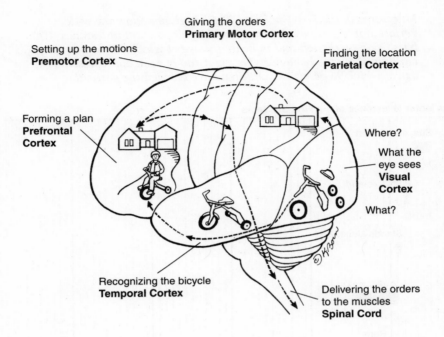

Giving the orders
Primary Motor Cortex

Setting up the motions
Premotor Cortex

Finding the location
Parietal Cortex

Forming a plan
**Prefrontal
Cortex**

Where?

What the
eye sees
**Visual
Cortex**

What?

Recognizing the bicycle
Temporal Cortex

Delivering the orders
to the muscles
Spinal Cord

mic pedaling, and the tires glide over the pavement. His *vestibular system* helps keep him in balance, and his cerebellum constantly monitors the movements of his muscles. The circle is complete. The boy's brain is performing its three main functions: collecting information, making sense of that information, and acting upon it. As we are left to ponder this, he happily disappears from our view.

Milestones of the First Six Years: The bar graphs show the wide variation in the times children reach developmental milestones. The bars begin on the left and indicate the age at which 25 percent of children show the behavior. The right end of each bar represents the age at which 95 percent of children have reached the particular

Gross Motor Milestones of the First Six Years

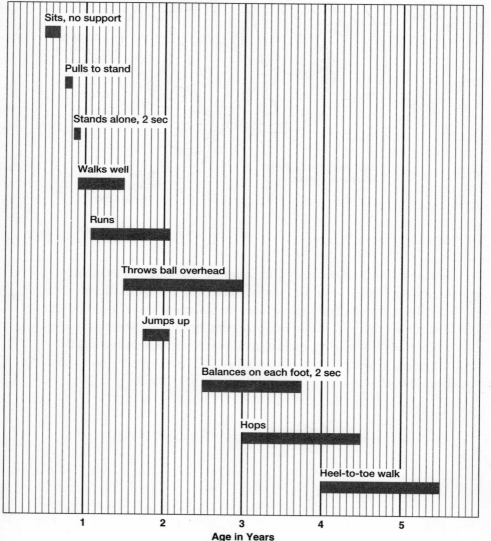

milestone. Adapted from W.K. Frankenburg et al. (1992) The Denver II: A major revision and restandardization of the Denver Developmental Screening Test, Pediatrics 89:91–97, and from M.D. Levine, W.B. Carey, A.C. Crocker, eds. (1999) Developmental-Behavioral Pediatrics, 3rd edition, Philadelphia: W.B. Saunders.

Fine Motor Milestones of the First Six Years

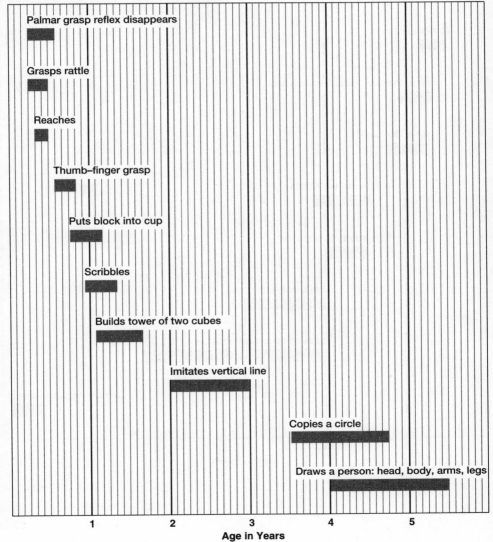

Language Milestones of the First Six Years

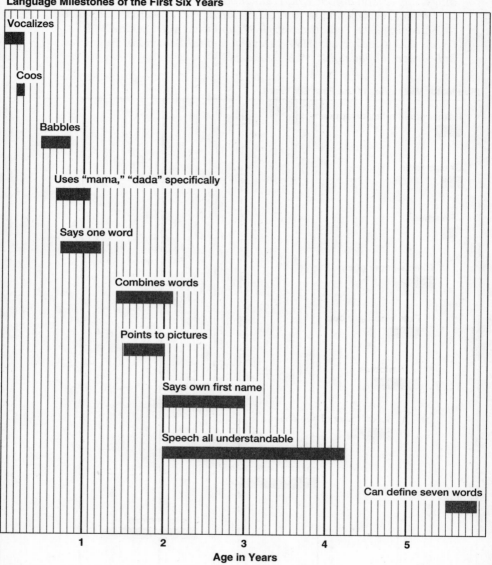

Vocalizes

Coos

Babbles

Uses "mama," "dada" specifically

Says one word

Combines words

Points to pictures

Says own first name

Speech all understandable

Can define seven words

1   2   3   4   5

Age in Years

Play and Daily Life Milestones of the First Six Years

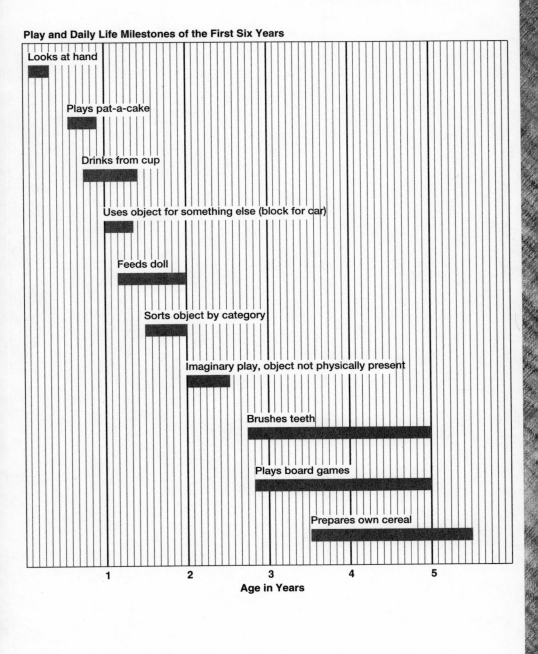

Part Three

The Second Year

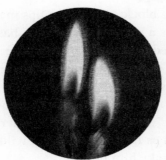

Two Candles for Emily

How different the party room looks from when we saw it one year ago! The scrambling infants have become toddlers who propel themselves around with rapid, short steps and eagerly explore every corner they can get into. Although they generally play side by side rather than together, they are becoming aware of themselves and others as persons and are beginning to discover the uses of language.

At the chime of the bell Emily dashes impatiently to the door to welcome Sonja and her mother. Her eyes light up as Sonja hands her a bright box tied up with a large ribbon. Matthew and Steven follow close behind Sonja. In a few minutes the room is filled with the sound of scurrying feet and toddler chatter.

Anna clings to her mother's side as her clear blue eyes anxiously scan the room. She pulls her thumb from her quivering lips and begins to cry. "Would it be all right if I stay?" her mother asks. Deborah tells her it's fine. Several of the other mothers will be staying too.

For a while, the children play happily on the floor. Matthew

pushes a large blue block around, his lips making the sound of a
motor. Sonja blows a whistle, and Steven has become fascinated
by the multicolored balloons hanging from the ceiling. Just as
Anna picks up the teddy bear lying in the corner, Tommy rushes
past her and pushes her away. She tumbles to the floor and begins
to cry. Emily hurries to comfort her friend. Steven mutters,
"Tommy . . . bad." When Tommy's mother scolds her son, he
screams and thrashes his arms and legs wildly. Deborah wishes
his mother would take him home.

Now it's time for cake and ice cream. Emily blows out her
two candles and then helps her mother pass pieces of cake around.
The children devote their full attention to eating the ice cream
with a spoon and drinking their juice from plastic cups. It isn't
long before the first children wriggle out of their seats and return
to the playthings waiting in the other room.

Deborah hardly hears the doorbell, the first of the mothers
coming to pick up the children. Tommy and his mother have already beaten a hasty retreat. Deborah sponges the sticky frosting
off Emily's face and begins to clean up the table. The other mothers help her pick up the room and take the dishes to the kitchen.
While they are doing that, let's follow the big "strides" the children have made between their first and second birthdays.

Discovering

Your toddler's world is rapidly expanding both outside and inside. Being able to get around and to manipulate utensils expands his possibilities for discovery. The growth of language abilities improves his lines of communication. On the inside, he is acquiring a sense of self-awareness, a growing memory capacity, and new ways of thinking, such as the ability to draw conclusions (inference). At the same time, he is becoming aware of the feelings and intents of others. The little explorer we saw at the beginning of the first year is now beginning to test the waters. Meanwhile, keeping track of what is going on in a toddler's brain to make all these things possible is like trying to keep up with a room of lively two-year-olds.

Getting Around and Getting Into

At Emily's first birthday party, most infant guests were crawling around on the floor, although a few children were standing or taking a hesitant step or two. Now, they are all walking. Two-year-old Matthew opens the door by himself and goes up and down the stairs, putting first one foot and then the other on the same step and using one hand on the wall to support himself. Steven can jump to try to catch the string of a balloon.

The reason the children are able to get around on their own so well is not only because their muscles have had time to grow. The part of the brain that tells the muscles what to do is becoming better able to do its job. Even in a toddler, it's a long way from his brain down to his toes. The signals have to travel down the long axons of the motor neurons in his spinal cord. These axons are now being surrounded by their insulating layers of myelin. Myelin eliminates interference from neighboring axons and prevents "short circuits" from taking place. This allows the signals from the motor cortex to travel more rapidly and efficiently to the leg and foot muscles. The improved efficiency of the spinal cord neurons also allows voluntary control of the sphincter and bladder muscles involved in learning to use the toilet. Myelination takes place sooner in girls than in boys, which—in addition to social factors—may help explain why girls are generally toilet trained earlier than boys.

The cerebellum is also important for walking movements. This structure receives its information from the motor cortex and from the spinal cord. Do intent and outcome agree? When the cerebellum detects deviations from the original plan, it informs the motor cortex to make the changes. Between the ages of one and two years, the nerve fibers connecting the cerebellum and motor cortex undergo a spurt in myelination, which greatly enhances their collaboration.

Handling a spoon

Just one year ago, Emily literally dug into her cake and ice cream, and not all of it made its way into her mouth. Now she uses her spoon to get her food on target. Being able to eat with a spoon is a big step on a child's way to mastering the skills of his culture.

Michael E. McCarty and his colleagues at the University of Massachusetts at Amherst studied the progress infants make in

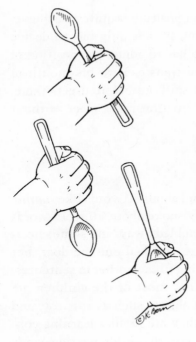

Grabbing the Spoon: It takes time and experience to grab the spoon the right way. (From M.E. McCarty et al., "Problem Solving in Infancy: The Emergence of an Action Plan," Developmental Psychology, 1999, 35(4), 1091–1101.)

learning to wield this handy utensil. The researchers placed a spoon on a simple rack in front of the child. Even some of the nine-month-olds managed to grasp the correct end of the spoon and manipulate the bowl into their mouths when the spoon was in the right position for them to grasp with the preferred hand. But when the spoon was placed in the other direction, they often grasped the bowl of the spoon and put the handle into their mouths. The 14-month-olds never put the handle into their mouths. However, they often used the wrong position initially to take hold of the spoon. They then either changed the position of their hands or switched hands midway so that the bowl of the spoon delivered its contents into the mouth. By the age of 19 months the children were able to block their impulse to reach out with their preferred hand. Instead they reached with the hand that was on the side of the spoon's handle in order to achieve the proper grip.

Grabbing the spoon in the proper position is quite an achievement. When an infant sees a spoon, he not only has to decide which hand is appropriate but also has to adjust the position of the hand to pick up the spoon and transfer it to his mouth in such a way that the contents don't spill out before it gets there. The impressive visual–motor coordination takes place without his actually "thinking" about it.

The Language Explosion

One-year-old Emily could understand about 50 words like *mama*, *dada*, *cake*, and *drink*, but she only spoke about 10 to 12 words herself. She and her little friends babbled away, and some cheerfully waved "bye-bye" as they were whisked out the door, but none were combining words or talking to each other in sentences.

At Emily's second birthday party, most of the children are using a number of single words for objects such as *cup*, *toy*, and *ball*. Their words also show that they are rapidly learning concepts such as "more" or "gone." Some of the children string words together into simple sentences, although it isn't always easy yet for people outside their family to understand them. Language is helping two-year-olds make social contacts and learn a lot about their world. But they also enjoy the "feel" of language, listening to the music of nursery rhymes and playing with the sounds of words. They enjoy associating an animal with the sound it makes and may for a while call a dog a *bow-wow* or a bird a *peep-peep*.

Visiting a country where nobody understands English will give you an idea of an infant's difficulties. If a cookie is on the table, a baby can point to it and signal his wish to have it. However, if the cookie is in the cupboard, his chances of getting it are brighter if he can ask for it. He can show his discomfort by making a face and crying, but being able to tell you "tummy . . . hurt" or "head . . . bump" will make it more likely that you undertake the appropriate action to remove the cause of his distress.

Language is so important for us that it is surely no coincidence that we use a language-related term to describe young mem-

bers of our species. The word *infant* comes from the Latin *in +
fans,* meaning nonspeaker. Becoming a "speaker" paves the way
into life with others. It is no wonder that we can hardly wait for
our children to start talking.

A young mother came to my office deeply concerned about
her 17-month-old son, Sammy. The boy had a sunny disposition,
smiled a lot, and looked attentively when she was speaking to
him, but he hadn't said a word. His mother was a language teacher
who had gone on record as an early talker herself, and she had
begun to worry about when her own child would start. I took
down the boy's medical history and did a complete checkup, including a hearing test. I asked the mother about the boy's developmental milestones, for example, cooing, babbling, and pointing.
Everything seemed in order. When I asked if he could understand
simple commands, she wasn't sure. So I told her to go home and
try the following "experiment."

She was to get her son's attention and speak to him slowly—
without any gestures or special intonation—saying, "Sammy, go
to your room and bring me your Teddy." This was his favorite toy,
and she knew that it was lying on the floor in Sammy's room. No
sooner had she spoken the words than the boy toddled off without
a moment's hesitation to get his bear. He had understood the
words. His mother was elated.

Although Sammy's mother was getting impatient waiting for
him to say his first word, little Sammy was not at all unusual.
There is a wide range in the times when children begin to speak.
It was most important to find out if Sammy could hear and understand what his mother was saying because this is a prerequisite
for speaking words. To his mother's great relief, a few weeks later
Sammy pointed to his toy duck and said "duck." From there his
vocabulary took off at high speed.

Babies understand words before they produce them. And they
always understand many more words than they actually use. Even
as adults we don't use all the words in our daily speech that we
understand when we are reading a text, listening to a lecture, or
watching a film.

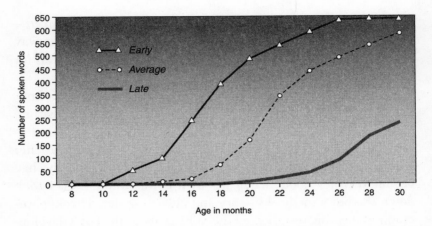

The Language Explosion: The graph shows the increase in the number of words spoken by an early, an average, and a late learner. Adapted from D.J. Thal et al. (1996).

Just as Adam in the Garden of Eden began to give names to all the animals around him, babies have a natural tendency to give names to things they see. The "naming game" is under way, although some cultures do not emphasize it as much as others. Learning and using new words is a process that usually begins slowly around the end of the first year and then goes into high gear around the middle of the second year. At around one year of age, babies have an average active vocabulary of about 10 to 15 words. By around 15 months this number has more than doubled. At the end of the second year, they may already speak around 400 words. And they understand about five times more words than they actually produce.

A baby's first words are closely tied to what is going on immediately around him. By combining a word with a gesture or facial expression, he can expand its meaning into a whole statement or demand. For example, "mama" spoken with outstretched arms means "I want you to pick me up," or "baba" (bottle) means, "I want my bottle." He can also use words such as "more," "done," "up," and "down" in this economical way. Parents quickly become experts in simultaneous translation, sizing up the situation, putting the sounds and gestures together, and filling in the missing parts of the message.

At about 20 months, Sammy suddenly began to put words together. Interestingly enough, this accomplishment depends more on the size of a baby's vocabulary than his age. It's as if a baby has to fill a storage room with 50 to 100 words before he can start combining them. Sammy started with "telegraphic" messages, such as "Dada . . . bye-bye," "Mummy . . . come," or "milk . . . gone." His vocabulary included a few verbs, mostly in the infinitive, as well as a smattering of adjectives or adverbs like *yummy, up,* and *done.* Around this time, he began to use his own name to refer to himself.

With the use of two-word combinations, Sammy was already showing that he had absorbed the idea of how words are put together in his native language. He put words together with subject and verb in the right order, for example, "boy . . . eat," and not the other way around.

Toward his second birthday, Sammy became better able to articulate the sounds that make up words. At first, he had been using the vowels and the "easy" consonants, like *b, g,* and *m.* Now he was adding new ones to his repertoire. This enabled him to make clearer distinctions between words like *stay* and *play* and to put the *s* on words to make them plural. However, people outside his family couldn't really understand what he was trying to say until he was about three years old, and two years later his first-grade teacher had to help him master the challenge of *th.*

The fact that the sequence of babbling, single words, word combinations, complex sentence structure is universal—with a few possible exceptions in which a child appears to leave out the single-word stage—is evidence that brain development underlies a child's acquisition of language.

Special language areas in the brain

The language areas of your child's brain are humming with activity. During the first three to five months, when babies are cooing and making simple melodic sounds with their lips and tongue, the part of the primary motor cortex of the right hemisphere that controls the muscles of the mouth and face is developing rapidly. But around 12 to 18 months, when the baby's sounds are becoming more precise and turning from babbling to first real words, the main building activity shifts to the corresponding areas in the left hemisphere. This newly active area of the motor cortex is very close to Broca's area, where intense development is also taking place. Broca's area is involved in the comprehension of speech sounds and is important for the articulation of words.

Broca's area is responsible for setting up the program for the sounds and sending it to the neurons in the primary motor cortex that control the movements of the mouth. Between 24 and 36 months, there is intense growth of dendrites in both the left and right Broca areas. At around four years the nerve cells in Broca's area will have settled into their final pattern of layers. This is the time when a child becomes able to pronounce words clearly enough for others to be able to understand what he is saying.

As children begin to put words together into sentences, another language area of the cortex steps into the foreground. Wernicke's area, named after the German psychiatrist and neurologist Karl Wernicke (1848–1905), receives information from the ears, eyes, and sense of touch, which is why it is so important for understanding all types of language, whether it is oral speech, sign language, or Braille. Wernicke's area is directly linked to the Broca center, which has connections to the motor centers controlling muscles involved in speech. The language network allows a child to understand speech, prepare an answer, and then speak the words.

In most people, language functions come to be mainly located in the left hemisphere. During the early years of childhood, when the basic centers are still under construction, damage caused to the left hemisphere leads to the formation of speech centers in the right hemisphere instead. This changeover illustrates plasticity, the amazing ability of the brain to compensate for lost functions by employing other areas. If the left hemisphere is damaged after puberty, some functions may still be taken over by the right hemisphere, but the period of highest flexibility is over.

The increasing specialization of the left hemisphere for processing speech is reflected in brain electrical activity. Dennis Molfese, at Southern Illinois University, gave parents of 16-month-old infants a list of common words, such as *bottle*, *cookie*, *key*, and *ball*, and asked them to mark the ones their babies knew. The parents then brought their infants to the laboratory. The investigators placed electrodes over each infant's scalp and seated the baby comfortably on a parent's lap. While the baby heard a series of "known" and "unknown" words over a loudspeaker, the electroencephalograph (EEG) recorded the pattern of electrical activity in his brain.

The children showed a distinctly different pattern of electrical activity depending on whether the word was familiar to them or not. The researchers also observed gender differences in the location where the distinctions were registered. When the boys recognized a word, the particular activity pattern was seen in both the right and left hemispheres. The girls were more likely to show the pattern only in the left hemisphere. Because the left hemisphere is usually the "mature" site of language functions, this could be evidence that girls are ahead of boys with respect to development in language areas in the brain.

Language gets a big boost from the many connections that are now being strengthened in the brain. During the second year, neurons in the cortex make their connections in the corpus callosum, which unites the two hemispheres. Now signals can travel back and forth, so the two hemispheres can pool their information. The right hemisphere processes the perception of objects and actions, the left hemisphere the actual names for them. Thanks to the improved linking of the two hemispheres, the information from both sources is integrated more rapidly and efficiently. Therefore, when the child sees a particular object, he can retrieve the word quickly and say it. This shows that despite a certain specialization of the two hemispheres, cooperation between them is essential.

Although a tremendous amount of activity is taking place in the language areas of the cortex, we should not forget the work going on "downstairs." The connections that reach downward from the cortex to the subcortical structures are getting stronger. Ties are being forged to the basal ganglia and cerebellum, full-fledged members of the language network.

Children's brains are programmed to process the necessary signals from their environment in order to set up basic language centers, and a child growing up with intact systems in a surrounding where words are spoken will automatically acquire speech as soon as those systems are ready. The timing and extent of language acquisition depend on both genetics and environment.

Hearing and using words

Because of the wide variation in the times when infants say their first words, it is impossible to say whether direct coaching will make a child start to speak sooner. However, speaking a lot to the child can make him more talkative. In Part 2 we mentioned the comparison of the infants in two neighboring African tribes, one whose mothers spoke a lot to them and one whose mothers spoke very little. Not surprisingly, the babies in the first tribe talked more. In cross-cultural comparisons, American mothers seem to be more active in animating their infants to speak than Guatemalan, Dutch, or Zambian mothers, and in general, American babies are more talkative. Janellen Huttenlocher, at the University of Chicago, compared the speech habits of mothers and their 16- to 20-month-old toddlers. When the investigators counted the total number of words used over a three-hour period, they were surprised at the differences. The least talkative mother used only 700 words, while the most talkative used almost 7,000. And sure enough, the most talkative mothers had the most talkative infants at 26 months! Does this mean that you have to keep up the nonstop patter of an early morning radio show? Luckily this is not the case. Other factors are more important than the sheer amount of verbiage a child hears.

When Lois Bloom and her colleagues, at Columbia University

Teachers' College, followed the language development of a group of children from the ages of nine months to two years, she noticed that two-thirds of the time it was the children who initiated the dialogue. Mothers were more likely to speak immediately *after* the child said something. They either repeated or acknowledged the child's words or attempted to clarify what the child had said. Bloom suggests that young children are actively involved in selecting the words they need from their environment themselves and that they use them with the intent to communicate what is important to them at the moment.

Marc Bornstein and his colleagues, at the National Institute of Child Health and Human Development, found that how mothers responded to their children was more predictive of the child's later vocabulary than the size of the vocabulary they used with their children. The investigators observed the babies first at 13 months, then later at 20 months, both at play and at mealtimes. At the same time, they noted how the mothers responded to what their babies said. When the researchers measured the babies' vocabularies at 20 months, they found a correlation between the mother's responsiveness and the baby's vocabulary. Mothers whose children showed the greatest increase in vocabulary were those who entered into the child's dialogue, sometimes repeating the child's words. For example, "the red ball. You want me to catch the red ball? Throw it to me." These mothers tailored their speech to their child's current level of language ability, gradually using more words as their child's language skills grew.

Direct interaction in conversation is extremely important for children's language learning. Radio and television by themselves have no significant effect on the vocabularies of very young children. However, if words or phrases are singled out by other viewers for particular attention, this may well encourage imitation.

Playing

Children's play reflects changes in the way they perceive and think about their world. While the one-year-olds were intent upon exploring their toys with both hands and mouth, the two-year-olds are now using their hands primarily; and they are using them more skillfully than they did as infants. They can put two cubes on top of each other to build a "tower," and they can manipulate a spoon. Two-year-olds know that objects are used for a specific purpose. Emily enjoys brushing her teeth with a toothbrush before she goes to bed. She knows that scissors are for cutting paper and that a broom is for sweeping the floor. She gives her bear an imaginary sip of milk from a cup.

When Steven was 10 months old, he banged two blocks together or put one in his mouth. He was listening to the sound they made or feeling their shape with his tongue. Four months later, Steven showed that he was looking at objects in a totally new way. He took a rectangular block of wood and pushed it along the floor, making his lips go *brrrrum . . . brrrum*. The block had become a "car." Using a piece of wood in place of a motor vehicle is an example of *inference*. Inference means making a kind of mental jump from one object or event to another. Steven's brain forms a temporary association between the block and the image of a car he has stored in his memory. The ability to make inferences and use mental pictures is the beginning of creative activity, which grows because children can now pay more attention to the distinctive features of an object and can store more memories over a longer time. Their brain is becoming capable of linking information more effectively, enabling them to make rapid associations.

Grouping objects

Around the middle of their second year, children become fascinated by grouping objects that look similar. Your toddler may want to put all the cups in one row and all the glasses in another. Sorting objects into categories is a practical way to store things so you can find them again. This is also the way the brain works. Forming imaginary groups to which items belong helps children to learn about them.

The ability to form categories becomes increasingly sophisticated during a child's second year. A group of investigators in England showed infants two pictures—for example, one of a typical bird, such as a robin or sparrow, and one of an unusual bird, such as an ostrich. The babies then heard the word *bird*. The 12-month-olds looked longer at the typical bird, but by 18 months, babies looked equally long at both examples of the category. They had expanded their category of "bird" to include examples that had more unusual features.

As children become more adept at detecting similarities and differences in objects and distinguishing the relevant features that form categories, they become better able to apply their information in new situations. Steven is intrigued by the wooden flute Emily's father has hanging on the wall next to the piano. He pushes the piano stool up close to it and clambers up so he can grasp the instrument and pull it down. He has never seen such a flute before, but the shape reminds him of instruments he has seen in a parade. So he infers that the flute can be played like the instruments he has seen before. He puts the mouthpiece to his lips and blows.

Just as children sort balls and blocks or fit shapes into the appropriate spaces in a shape board, they also sort other people into categories. They learn that they are members of a family, or play group, that they are "girls" or "boys." They also distinguish between "adults" and "children," or even between "children" and "babies" and have ideas about behavior that is appropriate for each category. Your two-year-old will probably laugh if he sees you drinking from a baby's bottle.

A simple little experiment shows that children can transfer categories about people to inanimate objects. Put a handful of small stones of different sizes onto the table in front of your two-year-old. Ask him to show you the "daddy stone," the "mommy stone," the "baby stone." See if he sorts the stones by size, making the biggest the "daddy" and the smallest the "baby."

The brain's built-in systems for sorting things into categories are a big help for language learning. Children intuitively sense that certain words are "names," and that others signify "actions," "position," or descriptive qualities. This enables them to build new combinations of words, depending on the situation. A toddler can say, "I see cake on plate," "I want cake," or "Mommy make big cake."

Cause and effect

Toward the end of their first year babies make the interesting discovery that their actions have an effect. When they drop the keys on the floor, they hear an interesting sound, and an attentive adult usually pops up immediately to return the keys. Both results of the key drop are fascinating for a while, and the action is likely to be repeated until the baby gets tired of it.

During the course of the second year, children set out more purposefully to see what happens. They have learned that a switch will turn on a light, that turning the knob on the radio will bring sounds, that turning the handle on the water faucet will let water into the sink. They want to confirm their knowledge at every opportunity. It's exciting to pull the plug in the bathtub, to drop a cookie into the toilet and then flush it. Trying to find out what will happen is often especially tempting after a parent has said, "Don't."

I will never forget one episode that occurred when we took our five-year-old son and almost three-year-old daughter on a short hike in the mountains. Before we started out, we warned the children not to touch the electric fences surrounding the cow pastures. We were walking single file ahead of the children, when we suddenly heard a sharp cry, followed by our son's explanation, "Jessica discovered electricity!" She hadn't been able to resist the temptation to see for herself what would happen when she touched the wire.

A two-year-old's drive to experiment with his environment doesn't only mean pulling levers and pushing buttons. He is also trying to find out what other people will do. He may say no to one item of clothing after another or answer with no when you ask him to pick up his toys. His no is not necessarily a sign of disapproval, refusal, or a generally negative attitude. Sometimes he just wants to see what happens when he says it.

Concentrating on a task

Sonja is concentrating so intently on fitting the wooden forms into the shape board that she doesn't even hear her mother call her to supper. She wants to finish the job. Sonja's "internal manager," located in the front part of her brain, helps her complete the job she has set herself. This central command headquarters performs a host of valuable functions. It helps her ward off distractions and focus her attention on the task at hand. At the right moment, it says, "Wait a minute. Let's try something else." When she succeeds at putting all the forms into the right places, the reward centers in her brain register the positive experience, making it more likely that she will enjoy doing things like this by herself in the future. It's the bonus she gets when her "manager" is pleased.

Sonja's command center is part of a team, and we can only highlight a few of the many participants on it. The prefrontal cortex helps her select the relevant details and keep them in mind. The frontal cortex, which extends further back and is closer to the motor areas, helps her plan her actions. News of her success travels swiftly between her cognitive centers and her limbic—that is, emotional—system. As her ability to understand and use language grows, it supports and enhances her capacity to direct her actions. She will be able to use words to tell herself what to do next. The toddler years are a time of intense construction in all of the networks involved in these processes.

The growing complexity of the brain networks provides the toddler with the basis for new activities, and in turn, the stimulation provided by the new activities encourages neural growth. Through experience Sonja learns to put her new abilities to use. This experience can be guided and encouraged by the adults around her.

How much help to provide in solving a task may depend in part on a child's age. Susan H. Landry and her colleagues observed how mothers and their children played together and, specifically, how the child responded to his mother's suggestions and how often the child requested assistance. The investigators then examined how children went about solving problems at four-and-a-half years old. How well could the children independently formulate problem-solving strategies and show flexibility in carrying out a plan? The researchers found that when the mothers were active in pointing out possible strategies to their two-year-olds, this guidance had a positive effect on the way the children went about their task at the age of four and a half. However, if the mother constantly gave the child instructions at three and a half, he was less capable of solving problems later by himself. The researchers concluded that parents should give fewer directions as the child becomes capable of developing effective strategies on his own.

To Think About

Can the reaching of motor milestones be speeded up?

Parents impatient for their children to start walking may wonder if training has any effect. Interesting observations in the !Kung San tribe in Africa help to answer this question. Because hunting is so important, the tribal adults made deliberate efforts to accelerate their infants' walking. They began before the infants could stand by holding them and helping them to make rudimentary walking steps. However, the training had no effect. The infants began to walk between 11 and 14 months. (On the other hand, in other cultures, close swaddling or the use of cradleboards does not delay walking.)

When will he finally speak?

Be patient. How familiar this must sound! But it really isn't easy when your neighbor's baby is blurting out whole sentences while your toddler hasn't produced anything you would honestly call a word. There is a wide range in normal language development. However, it is important to observe whether your child responds to your words in ways that show he understands. Does your two-year-old enjoy imitation games? Does he follow simple requests and use gestures to indicate his wishes? Does he consistently use a combination of sounds for a particular object, even if it doesn't sound like the real word? If he doesn't do these things, he may still just need time, but it would be better to discuss the question with your pediatrician.

If your baby has shown that he *is* able to use words to ask for things, give him a chance to use the word. Hesitate before handing him the piece of cake when he just points to it and makes *mmm . . . mmmm* sounds. If he uses a word that isn't clear, don't correct him. Just repeat the word correctly in your next sentence.

Once your baby has discovered the fun of using words, you can help him learn more of them. He will be more likely to take them over when they are related to what he is interested in or doing at the moment. Babies often enjoy it when you point to something they know and they can say the name. You can extend this to "What sound does the airplane make?" or "Make a happy face." Responding positively to your child encourages him to occupy himself for a longer time with an object, giving him more opportunity to associate it with the new word. Two-year-olds like to look at picture books together, which gives them a good opportunity to point out an item and ask you for its name.

One language or two?

You may be living in a bilingual environment, or perhaps you have a spouse, partner, or in-law who speaks a different language. This is a wonderful opportunity for your child to grow up with more than one language, to learn it naturally, in the same way he learns his native language. The ease with which young children learn two languages at the same time may be due to the fact that they are learning both languages with the same "first language" brain systems.

Young children automatically absorb the language they hear spoken around them. Although we live in an environment where people speak a Swiss dialect of German, we decided to keep English as a family language because that was the language of the children's mother. After spending a few days with our friends while we were at a meeting, our two-year-old pointed to his undershirt and proudly said, "He—meaning our friend—say 'Hemdli'," which means "little shirt" in Swiss German. Without any coaching on our part, the children made the distinction between people who spoke German and people who spoke English and insisted on using the appropriate language with them. Thirty years later the "children" still speak English when they are together.

Does he need a lot of toys?

No. Just as your bustling toddler needs room to exercise his growing muscles, he needs room to stretch his powers of imagination. And he can do this best with simple objects that he finds around him. I once saw a harried young mother with a two-year-old in tow on the path of a park in Basel. Her left side was bent over by the weight of a huge plastic bag, out of which were sticking the various multicolored appendages of numerous toys. Her son dragged a lightweight plastic car behind him. Was the mother afraid her son would be bored? He had the whole park before him, where he could watch the squirrels, dabble his fingers in the cool, clear water, build miniature dams and bridges out of sticks and stones, or sail leaf boats down the river. If the boy needed a bit of encouragement, his mother could have quietly begun to make her own bridge out of the shiny pebbles. Toddlers don't remain spectators for long.

When should I encourage toilet learning?

In their Guidelines for Health Visits, the American Academy of Pediatrics expects "some progress toward toilet training" by the age of three years. To be more precise, this means that by this time 90 percent of the children perform their bowel movements on the potty. Eighty-five percent are dry during the day, and 60 to 70 percent also during the night. Girls tend to be toilet trained somewhat earlier than boys.

A prerequisite for learning to use the toilet is that your child's nervous system has developed to a stage where he can control his sphincter and bladder muscles voluntarily. This means that the long connections from his motor cortex to the spinal cord neurons are myelinated. A child also has to make an association between a feeling in his body and impending urination or defecation. His brain has to be able to deliver its message to the sphincter and bladder muscles. If your child takes pride in doing things on his own and if he enjoys imitating you, learning to use the toilet makes him feel like one of the "big" people. It also helps if your child can identify body parts and is able to walk easily, pull off clothes, and get on and off the potty. Parents who want to nudge the process of toilet learning on a bit, can look for a pattern in a child's elimination times and get him to the toilet at an opportune moment. It may help to let a little water run in the sink.

Should I change my toddler's routine?

A story tells of a two-year-old whose nurse always bathed him in the same tub and in the same way. When she had to be away, a substitute bathed the child. The boy cried every time, causing the poor woman to wonder what she was doing wrong. When the old nurse returned, she asked the child, "Why did you always cry?" and the child replied, "She bathed me backwards." Where one nurse had started with his head, the other started with his feet.

Toddlers are beginning to understand that events happen in a particular sequence, so they often defend their sense of order. It's best to acknowledge their feelings and give them a short reason if you cannot comply with their wishes. However, they also have to learn to be flexible when situations change. If the new baby is hungry, his older brother may have to wait for his story time.

Me and You

The one-year-olds at Emily's first birthday party cruised around the floor with as little regard for their peers as if they were colorful obstacles blocking their path. To a two-year-old, other people, both children and grown-ups, have become a reality. Two-year-olds distinguish between *me* and *you* and are beginning to sense that others have desires and intents similar to their own. People can be sad, happy, angry, disappointed, surprised—all feelings that a child is also beginning to recognize in herself. Making the acquaintance of their own self and becoming an active participant in the social network are a large part of the discoveries children make during their toddler years.

It's Me!

"Self!" Sonja loudly protests, as her mother tries to help her put on her raincoat. This simple word reflects a truly monumental step that has taken place in her development. She has become aware of herself as a person. Without self-awareness, she would take no pleasure in her own accomplishment, nor would she feel responsible for what she does. She would not be capable of sharing the feelings of others, and she would not form lasting memories of events she experiences. However, a toddler's wondrous new sense of self can try the patience of the most easygoing parent, particularly when the child insists on putting her socks on by herself or when she tests the power of the word *no*.

Self-awareness makes its appearance during the course of the second year. In a classic set of experiments, children were first allowed to look at themselves in the mirror. They were then called by their mother, who unobtrusively marked their nose with a spot of rouge before they were allowed to look at themselves once again in the mirror. No child under one year of age showed any special reaction and touched his or her nose, despite the reflection of rouge on the face. However, there was a steep increase between 15 and 24 months in the proportion of children who touched their noses. This fact suggests that a child is beginning to infer that the reflection in the mirror with the rouged nose is herself.

These observations have led psychologists to suggest that during the second year a child develops an internal representation of how she looks and that this internal representation can be compared to the external mirror image, a process that involves working memory. Shortly after this competence emerges, children begin to use personal pronouns or their own name to refer to themselves. Other signs of growing self-awareness are a shy smile and averted gaze.

The appearance of self-awareness is related not only to mental images of the self but also to physical sensations. In Chapter 3, we talked about how some children reacted more intensely than others to their inoculations at 14 months. The researcher, Michael Lewis, observed that the children who reacted more strongly to their inoculations showed signs of self-awareness earlier than the low reactors. They also tended to experience more embarrassment as they got older. This finding indicates that a sensitivity to feelings of one's body is a component of self-awareness.

In Chapter 6 we saw that around the middle of their second year, children become very interested in sorting objects into categories. At the same time, they also begin to put people into categories, such as adults or children, male or female, family or nonfamily, and assign themselves to one or more of the paired categories. Between two and three years of age, children are aware that certain possessions and tasks belong to adults, and they are beginning to form gender-related stereotypes: boys are stronger, dads fix toys, mothers show more sympathy when you stub your toe.

Self-awareness adds new dimensions to social and emotional experiences, both positive and negative. Being aware that one doesn't have the toy that another child owns provides the basis for envy and jealousy but also for generosity. Knowing that you were the one who spilled the milk or broke a vase can lead to feelings of guilt, but knowing that you were the one who set the table can lead to pride and a growing sense of competence and responsibility.

Part of self-awareness is sensing what the self is capable of doing. Studies have shown that children begin to realize this around the middle of their second year. Eighteen-month-old infants were asked to imitate actions performed by an investigator, like building a stack of blocks or enacting a scene using blocks for animals. When the actions became more complex, many of the children began to fret and cling to their mothers. They sensed that the task was too difficult for them to perform.

The sense of self-awareness that children show as toddlers is still a long way from the conscious view of themselves that they will have as adults. However, it is the basis for developing habits and forming attitudes about their own competence. A child's belief in her own ability to achieve is a strong motivation for learning. A child whose experience has taught her to doubt that her efforts will meet with success is easily discouraged or bored.

How Others Think and Feel

Self-awareness is so closely entwined with an awareness of how other people think and feel that it is hard to discuss the two aspects separately. Knowing their own feelings and sensing these in others makes it possible for toddlers to enjoy a family joke, to share thoughts in a real conversation, to comfort a baby sister who fell and hurt her knee.

What does your child do or say when you accidentally bump your head on the shelf? If she comes to you looking sad or concerned and asking if you are hurt, you would say she is showing sympathy. She feels *for* you. Psychologists often use a broader term called *empathy*, which emphasizes the ability to participate in another person's feelings; in other words, to feel *with* them. From around the middle of their second year, children begin to actively show concern for others. They are no longer simply echoing another person's distress but indicating that new thinking abilities allow them to realize what has happened to the other person and to sense how that person feels about it.

Empathy often leads to a desire to help. By the end of their second year, most children actively enjoy imitating adults in their daily tasks. They do this so spontaneously and eagerly that their efforts go beyond mere imitation to a real wish to participate in accomplishing what the adults intend to do.

Depending on their temperament, children are very different in how they express empathy, and girls tend to show more empathy than boys do. However, showing empathy can be encouraged, both by the example you present in how you act toward others and by reminding your child of how she would feel under similar circumstances.

As important as it is to realize that others can feel the way we do, it's also essential to understand that other people quite often have feelings that are different from ours. During their second year, children begin to shift their point of reference from themselves to that of another person. One of the ways they show it is through their awareness of the desires of others.

Betty M. Repacholi and Alison Gopnik, at the University of California, Berkeley, tested the ability of 14 and 18-month-old children to infer an experimenter's preference for one kind of food over another. First, the experimenters found out that the children preferred crackers when they were given the choice between crackers and broccoli. This may come as no surprise to you, but experimenters have to be sure. Then one of the researchers tasted each of the foods and clearly showed disgust at the crackers and enjoyment of the broccoli. When the experimenters asked the children which food the taster might want on a further occasion, most of the 14-month-olds chose the food they preferred themselves, namely, the crackers. However, 87 percent of the 18-month-olds correctly guessed the taster's preference.

Around the same time that they become aware of other persons' desires, children form an idea of what another person is planning, or intending, to do. Andrew Meltzoff showed that even before they can speak children are able to sense what an adult demonstrator intends to do with an object, whether he is successful or not. For example, 18-month-old children saw an adult holding a miniature dumbbell consisting of a wooden cube on each end of a wooden dowel. The examiner tried to pull it apart. However, his hand slipped, putting an end to his attempt. When the children were handed the dumbbell, they successfully pulled the dumbbell apart, showing that they had formed an idea of the action the experimenter had intended to complete.

Evidence that young children are rapidly developing a sense of what others intend to do is provided by studies that show that even two-and-a-half-year-olds are already capable of employing strategies to prevent persons from reaching their intended goal. Michael Chandler and his colleagues at the University of British Columbia played a hide-and-seek board game in which children were encouraged to hide a "treasure" in one of several differently colored plastic containers. The experimenters gave the children a puppet called Tony and showed them how Tony's inky footsteps left tell-tale marks on the washable white playing surface. Then they told the children to have Tony hide the treasure so that someone who didn't see them do it wouldn't find it. The children relied on strategies such as erasing the footprints, lying, or even a combination of erasing the footprints and producing false trails, behavior clearly planned to mislead others about the true location of the treasure.

A child's discovery of the intents and assumptions of others can be a source of family amusement. Our own children made quite a game out of drinking my last sip of tea at breakfast. It was always the same: While I wasn't looking, one of them took my cup and drank the last teaspoonful of tea. Then I would turn around, saying how I was looking forward to that very special, very last sip of tea. And every time, I had to feign great surprise and disappointment, while they giggled with delight at their little joke.

A Budding Idea of Right and Wrong

When Tommy pushes Anna at Emily's party, little Steven immediately calls Tommy's action "bad." The children all sense that what Tommy does is not an accident, that it is "wrong."

The sense of right and wrong that begins to appear in the second year is related to all the other exciting developments that take place around this time—things like a child's increased ability to make inferences, a growing sense of self-awareness, and an ability to perceive and respond to the intentions and feelings of others. Children are aware of actions that meet with disapproval, whether those actions concern other persons or objects.

One of the ways to detect the emergence of a sense of right and wrong is to observe children's reactions to damaged objects. Psychologists have found that children have a tendency to focus attention on flawed, broken, or dirty objects. If the room is filled with a variety of toys, two-year-olds are surprisingly apt to single out the doll with the torn-off arm or the train with a missing wheel.

Grazyna Kochanska and her colleagues studied the relationship between toddlers' sensitivity to flawed objects and their concern about wrongdoing. In the first part of the investigation, the researchers observed children's reactions to pairs of objects, one whole and one damaged or broken. The two-and-a-half-year-olds definitely preferred the undamaged object. However, the flawed object attracted their attention. They saw the damage as something negative that had to be repaired.

In a second phase of the investigation, the researchers observed the children's reactions to damage they believed they had caused themselves. While they were playing with a "special" doll, the doll's head fell off, or when they picked up a T-shirt, a bottle of ink tipped over and stained it. (The investigators soon consoled the children by explaining the accident and producing the "real," unharmed, article.) The children who had been most adept at distinguishing between the damaged and undamaged objects in the first part of the experiment turned out to be the ones who showed more concern about causing the damage in the second set of experiments. Children who feel more uncomfortable about causing harm may be more attentive to signs of damage.

By the middle of their second year, most children begin to look for signs of approval or disapproval from adults. They become alert to parental standards and begin to qualify their own behavior with respect to the standard. They associate events with the words *good* or *bad*, and these words come to qualify actions that are promoted or discouraged.

Qualifying actions as good or bad has an emotional basis that is biologically related to how one feels. The child may associate the category of "punishable" acts with a tense feeling in the stomach or changes in breathing. The word *bad* is also used in connection with things that have an unpleasant taste or smell. The kind of associations that the child makes and the physical sensations that accompany them are influenced by the child's temperament. Children with highly sensitive nervous systems will feel greater agitation at the prospect of wrongdoing than children whose nervous systems react less strongly.

The ability to infer the cause of an event is essential for a sense of right and wrong. When a two-year-old child sees that a toy is broken, she infers that the damage was caused by someone's action, and she has learned from past experience to anticipate her parent's reaction. If she senses that she was the cause of the damage to the toy, she reacts with an expression of uncertainty or fear and may even say "uh-oh." Between the ages of one and two years, children may begin to show a desire to "repair" a situation if they have been the cause of the distress. By the time they are three years old, they do this more consistently. They are more capable than the younger children of making the link between the situation and what led to it.

The matter of respecting and protecting objects is pretty familiar to toddlers. How often they hear, "Don't touch that ashtray! Leave the bottle alone! Keep your greasy hands off the curtain!" Even if a house is pretty well childproofed, there are bound to be a few things around that are off-limits. Children often find these taboos highly intriguing. Judy Dunn, who studied the development of morality in young children, found that while children react with distress and arousal to events that depart from parental standards, these same events can also be the source of interest, amusement, and excitement. This observation isn't surprising when we think of the attention paid to crime and scandal in our newspapers, talk shows, and television news.

How well children cooperate with parental standards depends partly on their experiences with their caregivers. Studies showed that infants who at 14 months experienced more positive interactions with their mothers and whose mothers were more responsive were more likely at 22 months to show greater distress over a violation of standards and more eagerness to imitate the mother's actions. A spin-off of this behavior was that the mothers made less use of direct disciplinary measures. Chronic abuse—including neglect or extremely harsh punishment—may cause a child to categorize herself as "bad" and devoid of any capacity of doing good. As a result, she may show little sense of empathy or shame.

Toddlers are beginning to sense that their own actions have consequences, and they are capable of learning to inhibit an impulse that could cause harm or distress to another person. A feeling that things can be "right" or "wrong" comes naturally to them. However, learning the specifics of right and wrong depends on the guidance of the adults around them.

Linking the Hemispheres

When we are trying to come to a difficult decision, we often say that two heads are better than one. This is the way the brain works. The two hemispheres are not exactly identical. Although they share most features, each hemisphere has its own special area of expertise. The specialization of the two hemispheres is more prominent in human beings than in other species. By linking its two hemispheres, the brain has access to vast resources. It's a tremendously efficient way of doing things.

Perhaps the biggest brain news of the second year is the strengthening of connections between the two hemispheres. Although the primary connections—a bundle of around 300 million axons forming the bridge called the *corpus callosum*—were laid down shortly before birth, a lot of intense building activity takes place on both ends of the bridge during the child's second year. In the cortex of both hemispheres the nerve cells grow longer dendrites to make synaptic contacts with axons arriving from corresponding areas in the opposite hemisphere. This linking up allows a dramatic increase in the convergence of information from the two hemispheres.

The many new psychological abilities and behaviors that emerge during the second year are related to the increased integration of the two hemispheres. We have seen how this works for language. Thanks to the improved connections between the hemispheres, when a child sees an object, she can rapidly retrieve the word to go with it. The language explosion takes off.

In a similar way, the integration of information plays a role in the emergence of self-awareness. The right hemisphere processes the sensory feelings of the child's body. The left hemisphere contains the child's name and the words for her feelings. This also has an effect on a child's growing sense of right and wrong. When she has done something she feels is a violation of parental standards, her right hemisphere contributes an awareness of her emotional reactions, while her left hemisphere adds the labels *right*, *wrong*, *good*, or *bad*.

The three emerging competences (language, self-awareness, and standards of right and wrong) all require the capacity for inference. Inference gets a major boost when the prefrontal cortex adds its ability to associate information from different cortical areas. This is the time of rapid development in the prefrontal cortex, which will play a starring role in Part 4 as the major coordinator of such higher brain functions as memory, learning, planning and carrying out actions, solving problems, and making judgments.

Toddlers and Temperament

In Chapter 3 we described how the Harvard Infant Study team is following children's reactions to novelty over the long term. At four months, babies like Emily placidly sit through procedures such as listening to a strange voice or having a mobile waved unusually close to their eyes. These same unfamiliar events are very upsetting for babies like Anna. Emily and Anna represent the two extremes: children who react least to the unfamiliar situation (uninhibited) and children who react most (inhibited). Let's pretend again that our model children are taking part in the Harvard Infant Study and join the researchers as they observe the girls again at around 20 months, this time to see how they react to being placed in an unfamiliar room.

On unfamiliar territory

Together with her mother, Emily enters a room she has never seen before. The room is filled with strange and exciting things: a sloping floor, a box with a spooky dark hole to crawl into, a scary mask on the wall. Emily immediately leaves her mother and bounds across the room, laughing and obviously enjoying herself. She waves her hand into the mouth of the mask, crawls inside the "cave," and runs up and down over the sloping floor. When Anna enters the same room, she keeps her eyes fixed on the floor and clings to her mother. After the briefest glance at the unfamiliar room, her features darken with fear. She pushes desperately with all her strength against her mother's side to force her out of this strange room as quickly as possible.

The Harvard team found that those children who, like Anna, had shown great distress and high motor activity at 4 months of age were more likely to be fearful and shy at 20 months. The children like Emily, who had been complacent during the experiment at age four months and who had smiled a lot, were less likely to be fearful.

Previously we saw that Emily's and Anna's nervous systems had different "presettings," which affected their immediate reactions to the strangeness of the distorted voice and the closeness of the objects dangling from the mobile. Now that the children are older, the changes taking place in their nervous system are adding further dimensions to the way they react to unfamiliar situations. A child is beginning to interpret the situation in her own way. She is better able to recall previous experiences and compare them with the present situation. She is also more aware of the feedback her brain receives from her internal organs.

Anna feels her discomfort intensely, and she may associate these unpleasant sensations with similar situations in the past. She wants to avoid them if possible. Emily, on the contrary, may have experienced a good feeling when she was able to explore new territory on her own, so she sees this visit as an opportunity for discovery.

Using EEG as a window on basic moods

One characteristic that belongs to temperament is a child's basic mood, which can be related to how a child approaches novelty. Electroencephalographic (EEG) studies provide a special kind of window for observing directly in the brain whether a child shows a more positive or more negative attitude. They also show that parts of the frontal cortex are very much involved in the processing of emotions. When we talked about the appearance of separation fear during the first year, we mentioned that differences in EEGs predicted which infants would begin to cry sooner when their mothers left the room. Those who showed relatively more right frontal activation were the ones who cried earlier. As children get older, more and more interesting correlations between EEG patterns and behavior come to light.

Richard J. Davidson and his colleagues at the University of Wisconsin used EEG techniques to study the differences between outgoing (uninhibited) and shy (inhibited) children like Emily and Anna. They observed 31-month-old children playing together. Two children and their mothers were in the room at one time. The investigators noted behaviors such as how far the children strayed from their mothers, how long it took the children to begin speaking, how long it took them to enter a toy tunnel or to touch toys left on a tray by a stranger.

On the basis of these observations, the researchers described the children as "uninhibited," "inhibited," or "intermediate." When the investigators measured the baseline EEGs of the children a few months later, they found that the uninhibited children showed more left frontal activation and the inhibited children more right frontal activation, corresponding to earlier findings on separation fear at 10 months.

Putting together the knowledge gained from different types of research in the field of temperament will help us understand why children don't all react the same way to the same situation. Just as we are beginning to find correlations between high reactivity, shyness, and right frontal activation, we may also find new insights to help explain why some children are impatient and others persistent and why some get upset or angry quickly while others have a tendency to "stay cool." However, as a child grows, her experiences in daily life come to exert an ever stronger influence in shaping her behavior.

To Think About

What shall I do when she's testing my patience?

Learning about herself and others is an exciting part of a toddler's life. It is no wonder that she wants to find out what happens when she says "no," wants to exercise her ability to make choices, wants to see how far she can go. What a two-year-old wants to do and what she can do are often "out of sync"—a condition not unfamiliar even to adults. However, they have usually learned how to cope with it.

Your job as a parent is to help your child find her way in a manner that encourages both her ability to make her own decisions and her ability to understand her limits, whether these are the needs of other people or the constraints of the situation. Two-year-olds are just beginning to get the idea, so they need clear signals and brief explanations, like "Don't pull my hair. It hurts me."

Nothing can set off your own stress system like being in charge of a screeching toddler in a supermarket. If a stern expression and a swift, clear "no" are ineffective, your two-year-old will not be convinced by any of your arguments. If distracting her does not help, leave the shop and head for a quiet place outside where she can calm down.

It's better to reduce the chances that a child will get into the habit of using temper tantrums to get what she wants. Although most children will lose their temper now and then, the ones who do so frequently have learned that it is a strategy that works. Don't give in and reward a temper tantrum. Ignore it if possible, or play it down. Many everyday situations offer an opportunity to set patterns that make outbursts less likely. Children should learn that they don't automatically get everything they want when they want it. It is much harder for a child to learn the ropes when the rules are always changing. So try to be consistent. Parents should make every effort to set clear limits, keep their promises, and avoid empty threats.

How can I help my over-anxious child?

Parents of children like Anna may be tempted to structure the environment to avoid situations that are stressful to their child's temperament. However, since she will probably soon be confronted with unfamiliar situations and new people when she goes to day care center or nursery school, it is better to prepare her for these situations. This means allowing her more time to get to know her baby-sitter before you go out, going to the doctor's office early enough to allow her some time to play in the waiting room, or asking friends to be more patient and not approach her too suddenly or energetically.

An authoritative—in contrast to a permissive—parenting style has been shown to be particularly beneficial for extremely fearful children. This means fostering a feeling of security by providing guidelines for the child's behavior. It might mean, for example, making your desk "off-limits." If your child heads in that direction after you have said no, pick her up and take her back to her toys. If she keeps calling you after you have put her to bed at night, reassure her that you are there and that everything is all right, but make it clear that this is bedtime.

The investigations by Richard J. Davidson and his colleagues that we mentioned in the section on using EEG as a window on moods led to practical considerations for parents. The study showed that the infants who were more inhibited and fearful at 31 months showed more right frontal activation than left. However, the authors say that this could mean two different things, depending upon whether the right activity was especially strong or the left especially weak. In the first case, a child could have a strong tendency to fear novelty and avoid it if possible. In the second, the child could be showing a lack of positive interest in exploring the new object. It may, therefore, be just as important to encourage a child to find novel experiences rewarding as to help him overcome his fear of them.

Terrible or terrific?

The twos have been described as both. Understanding the developments that go on in two-year-olds and being aware of the wonderful achievements of this age will help you over the challenges. A toddler's growing capability for inference may lead her to draw the conclusion that a temper tantrum gets her what she wants. But she can also learn that it doesn't. She may tire you out because she's everywhere at once. Think of her activity as an unbounded desire to explore, coupled with the new ability to get around on her own. When she insists on putting on a sock by herself, no matter how long it takes, think of her action as a sign of a blossoming sense of self. Thanks to her new capability of empathy, she may comfort you when you have hurt yourself. She will share in your little jokes. And she will delight you with each new word.

Part Four

Three to Six Years

Six Candles for Emily

"You're not going to leave me alone with them, are you?" Allen says with a groan, as Deborah dashes out the door to take Andrew to a doctor's appointment. "It's OK," she reassures him. "Emily helped me prepare for the party, and she knows where things are."

If you ask how the children have changed most since Emily's second birthday party, her father would say they're more like "people." Instead of playing independently side by side, they play together. They carry on not only a dialogue, but a real conversation. Different as they are, the children have much in common: they are learning to get along with a wider circle of people and acquiring the skills they need in their society. If, as toddlers, our little explorers were rowing about and testing the waters, now they are getting the hang of steering their own boats, getting ready to leave the harbor of total parental control.

Though Allen can't keep a tune in the shower, he does his best to lead the group in "Happy Birthday." Emily blows out her candles and distributes generous pieces of the cake. The children are soon engaged in lively conversation. The big topic is the first day of school coming up in September. Matthew says he wants to go to school because he plans to be a lawyer like his father. Matthew is a confident, outgoing boy, who already speaks in a very persuasive manner. Tommy says he'd rather play football. Anna listens without participating in the discussion. When she accidentally bumps the table with her elbow, spilling a few drops of her bright red punch on the carpet, she squirms uncomfortably, torn between feeling guilty and not wanting to call attention to her mishap.

"How about a game of softball in the empty lot next door?" Allen suggests. Emily jumps right up and says she'll be the captain of the first team. When Anna protests that she's never played before, a lively discussion ensues about whether it would be fair to give her an extra chance to hit the ball. The children grab the ball and bat and head out the door, calling "I'm the pitcher," "You're on my team," and "Tommy isn't allowed to kick or push." Allen will be umpire. He's actually beginning to enjoy this party.

Gaining Competence

Approaching their first day at school, Emily and her friends all feel very important and grown up. Throughout the whole world societies generally recognize that around the age of six years, children reach a stage in their development that enables them to begin systematic training in the skills of their culture and to gradually assume more responsibility for their own actions and for the family. Between the ages of two and six many changes go on in the brain that make the children ready to coordinate the movements of their muscles and to start developing the thinking skills needed for their future lives.

Mobility and Dexterity

When Matthew got a shiny red bicycle for his sixth birthday, he could hardly wait to try it out. Since he had already ridden on a tricycle, he knew how to pedal it, but he now had to learn to keep his balance and steer at the same time. His mother held her breath as he swerved wildly back and forth for a few yards, but then he headed off smoothly down the driveway. Matthew's little excursion gives us a chance to see what his brain can now help him to do. His ability to pedal his bike is not only a matter of getting signals from his brain to his feet. Now the long-range connections that link the specialized areas of his cortex are being strengthened, thanks to their myelin sheath. Communication between them becomes more rapid and efficient. Matthew's prefrontal cortex forms the intent to propel his bicycle forward. The prefrontal cortex calls up the necessary information from the visual cortex in the back of the brain to tell him about his location so he can steer to the right or left. A special area of his motor cortex, his premotor cortex, sets up the program for the sequence of his movements. It orders Matthew's primary motor cortex to send the appropriate signals to the motor neurons of his spinal cord. Finally, the axons of these neurons convey the message to the muscles. Matthew presses his feet on the pedals, and off he goes.

For his excursion down the driveway, Matthew's cortex needs the cooperation of his subcortical structures. Around this time, the connections between these structures and his cortex are also becoming stronger. Like an internal gyroscope, his vestibular system keeps him in balance so he doesn't fall off the bicycle. His cerebellum watches over his movements and constantly adjusts their programming. The cerebellum is an important part of the network involved in procedural learning, the kind of learning that takes place when a child practices an activity. Matthew was two and a half when he climbed up on the seat of a tricycle for the first time. His feet dangled over the pedals. He pushed down on them, but not with the necessary pressure. One pedal spun around, and he didn't go anywhere. It was only after a few tries that he was alternating his legs and pushing down correctly. From then on there was no stopping him. Research has shown that training modifies cerebellar circuits. Once a pattern is established, the memories of it are very durable. You can see this for yourself. Even if you haven't been on a bicycle for years, your feet will start making the correct movements as soon as you get back on the seat.

We can't speak of Matthew's pedaling without mentioning his basal ganglia, a group of brain structures that play a central role in motor activity. The basal ganglia are "well connected." They receive input from the whole cortex, meaning that they are involved in higher order aspects of motor control such as planning. They are, however, also instrumental in establishing unconscious control of learned movements. The two-way connections between the cerebral cortex and the basal ganglia undergo myelination between the ages of three and six years. These connections are very important when a child learns a skill such as writing. He first consciously thinks through the motions, which will gradually become automatic.

The basal ganglia have special receptors for the neurotransmitter dopamine, a chemical messenger that is especially important for controlling voluntary movements. You may have heard that a lack of dopamine is a main reason for the movement difficulties of patients with Parkinson's disease. When a child is around one to two years old, dopamine receptors rapidly build up on the dendrites of the neurons in the basal ganglia, and dopamine levels in the body increase until around the age of 10 years. Dopamine also plays a role in the pleasure experienced during an activity. This might explain why children take such delight in running, jumping, or pedaling a bicycle.

From Scribbling to Writing

Give a three-year-old a piece of paper and a crayon and let him draw a picture. Chances are you have trouble guessing what it is unless he tells you. For him, scribbling means mainly the pleasure of making the crayon move over the paper. But during the course of the next year, he will begin to portray objects or people with simple, stereotypical forms. A sun is a circle with lines indicating sunrays. A flower can be a circle surrounded by a tight row of little circles. A person has a circle for a head and two lines underneath for legs.

Preschool and kindergarten children are not only fine-tuning their thinking tools: their senses are becoming more precise, their hand–eye coordination is improving, and they can use their fingers with greater precision. By six years of age, most of them can oppose a thumb to each finger sequentially and can hold a pencil with a tripod grasp, a prerequisite for controlling a pencil in writing or drawing. The ability to use hands to make precise movements is especially developed in human beings. We have a longer thumb than apes do, and our thumb joint is more flexible. In addition, the relative size of our premotor cortex, which is specialized for programming the sequence of steps that go into making voluntary complex movements, is six times larger.

However, children's fine motor abilities show a wide range of developmental timing, and for some, handwork is a real challenge. It was for me, anyway. It may sound surprising to you, but when I went to primary school, way back in the 1930s, knitting was taught to both boys and girls. It was considered a means of training hand and eye coordination, with the intent of fostering the joy of creating an "artistic" and perhaps even useful product as well. We had the task of making woolen washcloths for the soldiers in our army. I struggled bravely with my knitting needles, but my woolen washcloth turned out to be a grimy gray on one end and a shiny white on the end my teacher had completed for me. What for me was the result of days of hard work was a matter of a few minutes for my teacher. I often wondered what the soldier thought of my two-tone washcloth.

In contrast to me, Emily's friend Sonja is particularly adept at using her hands for precision movements. When she was about three and a half years old, her parents, both musicians themselves, gave her a small violin and sent her to a colleague who specialized in early music education. Sonja was thrilled. She would furrow her brows and fit her fingers carefully over the strings, listening intently as she drew her bow across them. At age six she finds the correct position for her fingers more quickly. The fine control of her finger movements is made possible by the improvement of connections between her motor cortex and the muscles of her hands and by the collaboration of the other cortical areas.

Although the neurons of the motor cortex show a general increase in the number of dendrites they grow, practice in using the hands can stimulate this development. As a result, areas of the motor cortex responsible for the movements of the specific muscles used in training grow larger, evidence that the brain is shaped by experience. The result of investigations comparing professional violinists who began training early in life with those who began after they were 20 provides an example of this phenomenon. Those who began their training early showed a larger area responsible for movements of the left little finger in the motor cortex. An important development going on in children's brains is the linking of the two brain hemispheres. We have already talked about its significance for language and the emergence of self-awareness. The strengthening of the corpus callosum, the bridge between the two hemispheres, also means that movements of both hands can be more rapidly coordinated.

The strengthening of the connections between the cortex and its subcortical associates, the cerebellum and basal ganglia, plays a role in training precise movements until they become unconscious. When we talked about how Matthew learned to pedal a bicycle, we mentioned procedural learning, learning by actually performing a sequence of movements. Learning to hold a pencil or crayon to write and draw is also procedural learning. Voluntary movements are practiced over and over again until they can be performed smoothly and automatically. Once learned, they are remarkably stable.

Handedness is in the brain

Until well into the second half of the twentieth century, many educators felt that schoolchildren with a tendency to use the left hand should be trained to use the right hand. A few years later, other experts, with equally little evidence, countered that training a left-handed child to use his right hand could cause reading deficits.

Handedness has a basis in the brain and seems to have been present since early in human evolution. Apes tend to prefer one hand to the other, depending on the purpose. They prefer the left hand for reaching, and a mother ape automatically reaches out with her left hand to pick up her baby in an emergency. Apes prefer to use the right hand to manipulate objects.

Left-handedness occurs in all cultures of the world, although countries arrive at different estimates of the prevalence. It is interesting that an estimate for the United States shows that about 10 percent of the total population write with the left hand, while an estimate for Korea reports 1 percent. This difference is most likely due to different cultural attitudes with respect to training all children to use the right hand.

Describing handedness, as with most human characteristics, does not mean distributing people into imaginary boxes called left-handed or right-handed: if we were to draw a line with one end point labeled *left* and the other labeled *right*, we would find that people are distributed all along the line, with more located over on the right. The distinction is particularly difficult because people who write with one hand often use the other for other activities such as using a knife or throwing a ball.

You may remember that 15-week-old fetuses already sucked more on their right thumb and that the newborn's stepping reflex more often led off with the right foot. Between the ages of 18 and 34 months, children often use both hands, but they show an increasing tendency to use one hand more than the other. By the age of three years, 86 percent of all children use the right hand more than the left. Paralleling their hand preference, children between the ages of 18 and 36 months develop a tendency to prefer one foot to the other to initiate walking.

Between the ages of two and four years, the neurons in the primary cortex responsible for movements of the hand suddenly sprout a large number of dendrites. In right-handers, this spurt takes place in the left hemisphere, and in left-handers in the right. Since the left motor cortex controls the right side of the body and the right motor cortex the left side, this means that right- or left-handedness becomes more pronounced during this time.

Genetics play a role in handedness. However, there is no reliable evidence to support attractive theories linking handedness with particular intellectual or artistic abilities. Careful behavioral studies combined with modern brain scanning techniques such as positron emission tomography (PET) and functional magnetic resonance imaging (fMRI) and genetic models will provide us in the coming decades with more knowledge on the fascinating question of handedness.

Executive Functions

The words *executive functions* call to mind the venerable CEO of a large corporation presiding over a meeting of his firm's senior executives as they plan the budget and lay out a marketing strategy for the coming years. In a way, the brains of Emily and her friends are doing something similar. Both long-term tasks, such as preparing a show-and-tell session, and short-term problem solving, deciding which move to make in a board game, involve a group of steps similar to the ones faced by the CEO: deciding on a goal, setting up the steps to reach it, forming a strategy for overcoming obstacles, monitoring performance, evaluating the results, and learning from them for the future. The prefrontal cortex is the brain's CEO.

One of the main functions of the prefrontal cortex is working memory, the ability to retrieve a "picture" of a situation from its location in the long-term memory store and hold it online so we can compare it to the present situation. Neuroscientist Patricia Goldman-Rakic has suggested that working memory gives human beings the ability to guide behavior using ideas and thoughts, rather than just reacting to immediate cues.

Planning and adjusting plans

You probably wouldn't leave the planning of a party or a picnic in the hands of a three-year-old, but by the time a child is six, you know that she likes to be actively involved in the preparation. She can help make a list of children to invite to her birthday party, suggest games to play, and decide what kind of food to serve her friends. As a toddler, she tried out ways to solve a problem she saw right at the moment: how to get her father to buy her the ice cream cone, how to reach the raisin cookie you left on the shelf. Now, as a big six-year-old, she can visualize a future event and carry out actions to ensure its success.

A few weeks ago Emily volunteered for her turn to prepare a show-and-tell session for her kindergarten class. She decided to tell her class about her garden, so she thought about the various things she would need to take with her. She could take a flower-pot with a sprouting sunflower plant, a packet of fertilizer, and some flower seeds. Then she would be able to explain to the class what a plant needs to grow: sunlight, water, earth—and if it's to grow especially well, care and attention.

The big day arrived, and Emily could hardly wait to stand in front of the class. At first, her classmates watched her eagerly, but after a few minutes, when she was explaining the fertilizer, she could sense that their attention was drifting off. They began to wiggle around and chat with each other. Emily decided to change her plans and bring out the peanuts she had been growing. When the children saw the real peanuts dangling from the roots, they got all excited. Emily was pleased.

Emily was able to see her demonstration as part of a whole project, the series of interesting demonstrations by all the children in her class. She was able to prepare for it in advance, and she was able to switch her "strategy," or initial plan, when she saw that her friends were losing interest. She felt her project was a success.

What a difference Emily's project is from two-year-old Sonja's efforts to fit the wooden geometric shapes of her puzzle into her shape board. Sonja's "manager," her prefrontal cortex, helped her focus her attention on her task and keep her goal in mind, but her problem was right in front of her, and she solved it mostly by trial and error. Emily's control functions have been upgraded to a more efficient and versatile module with a lot of great new features.

Attention opens the gates

The five-year-old son of one of the lab technicians at our hospital frequently had to wait around for his mother to finish her work. While he was waiting alone outside the laboratory, he marched up and down the corridors noting the names on the doors and the numbers of the rooms. Little Christopher had selected this particular feature out of the multitude of competing impressions in his environment and concentrated on his task long enough to store the information in his memory. Within a few weeks, he was able to recite all the names and the corresponding room numbers and could proudly deliver messages to the right "address."

The gateway to solving problems and performing tasks is attention, a delicate balance of two simultaneous processes. A child has to be able to select and focus on what is relevant in a particular situation. He has to be alert and receptive and at the same time be able to filter out, or actively ignore, all the irrelevant background stimulation. And he has to keep this up for a reasonable amount of time.

Attention is the job of a complex network involving many brain structures. The prefrontal cortex plays a supervisory role. That's as we would expect, because we have already seen that the prefrontal cortex is important for working memory, the ability to keep memories online while they are being used. The prefrontal cortex regulates the interaction of the other components of the executive network.

Crucial for attention and working memory is the delicate interplay of various neurotransmitters. Some tell the neurons to fire, while others dampen them. If we think of the neurons as members of an orchestra, the neurotransmitters would be the signals the conductor gives to the musicians. For example, he may tell the drums to stop immediately, the violins to come in gradually. If he doesn't send the right messages, the drums might take over or the violins fail to show up at all.

Dopamine is one of the messengers that modulate the activity of neurons. A disturbance in dopamine function can lead to attention deficit hyperactivity disorder (ADHD). Another important neurotransmitter is called neuropeptide Y, or NPY. NPY is found mainly in neurons that form small local circuits and play a role in the balance between exciting and dampening. They are involved in emotions and working memory, thus forming an important link between thoughts and emotions. The amount of NPY increases sharply between the ages of four and seven years, and the adult pattern of NPY is reached quite young, between 8 and 10.

"Elementary, my dear Watson": observation and deduction

By the time they are four to five years old, most children assume there is a reason for everything, and they want to know it. Their profusion of questions shows how their ways of thinking are changing as they grow older. Toddlers make automatic associations between events. They may have learned that a happy smile accompanied by a "please" will help get them a balloon or a cookie. Four-year-olds not only have more memories in their store, they are also better able than toddlers to apply their acquired knowledge to new situations. They begin to use their capability for deduction, a higher form of linking associations than the kind of inference that takes place automatically in the minds of toddlers. A four-year-old knows that a picture book costs money. He can figure out that three picture books will cost more than just one book. He may ask why books cost money.

Children not only ask questions. They also begin to make use of their own observations to investigate cause and effect systematically. A child might put a nail into a glass of water to see if it will rust, put a block of wood into the lake to see if it will float, or squeeze an egg to see if it will break. If you're lucky, he will ask you first.

A child's thirst for knowledge and wondrously unfettered imagination never cease to astound us. Educators dream of finding the magic key to preserve this enthusiasm and curiosity through all the school years and, indeed, through life. However, as Howard Gardner points out in his book *The Unschooled Mind*, "preschool thinking" has basic limitations. The tendency to think in stereotypes and be satisfied with simple explanations is understandable and useful as a substitute for a child's limited experience. But a child has to outgrow it if he is to make the step from acquiring knowledge to real understanding.

Strategies

The word strategy doesn't have to evoke images of armies deploying their troops or of advertisers eagerly aiming a product at consumers. It can be used to describe any systematic plan of action to reach a goal, involving even something as basic as moving one's eyes to scan an object.

Elaine Vurpillot measured children's eye movements as they performed a simple visual task. She gave the children two line drawings of a house and asked if they were identical. The younger children, average age two and a half, scanned the pictures haphazardly and made many mistakes. The six-year-olds, however, moved their eyes systematically from the top row of windows in each picture down to the doors. They were actively employing a "strategy" to help them in their task, and this enabled them to determine better if the houses were identical. They made fewer mistakes.

Maybe you have once relaxed with a page of games that magazines often include for the amusement of their readers. Did you take a pencil—or more boldly—a pen, to work out the maze? William Gardner and Barbara Rogoff showed a simple form of one of these mazes to a group of young children. The three-year-olds just "jumped in" with their pencils, but the four- and five-year-olds roamed the maze first with their eyes, testing the paths to find the correct one.

Strategies take on particular importance as children enter school. Robert S. Siegler, who has done comprehensive analyses of children's thinking, tells us that a child often tries out a variety of strategies and ways of thinking, rather than just a single one, to solve a given problem. The diverse strategies coexist until the child has had enough experience to decide which one works best for him in a particular situation. It is important to note that discovery can result from both success and failure. A child's mistake can frequently serve as an opportunity to bring about a change in the way he approaches a particular task.

Early variability in using strategies is related to later learning. A child who uses a greater number of strategies and who frequently corrects himself as he tries to solve a problem is better able to acquire new knowledge. Siegler suggests that a possible way to encourage children to develop their strategies is to ask a child to explain both why correct answers are correct and why incorrect answers are incorrect. This helps focus attention not only on the result but on the path the child took to get there.

Between the ages of three and six years, children make frequent use of the strategy of talking to themselves, or using "private speech," to direct their behavior as they work things out. Private speech serves a number of important functions. It helps the child organize his thoughts. It focuses his attention on what he is doing. By talking to himself, a child can go through the appropriate steps and make the necessary changes in order to accomplish his task. Through repetition, the steps become internalized. Private speech can even serve as the child's own "cheerleader," telling him to keep going, try something else, or like the famous little locomotive, convince him that "I think I can." Once a task is mastered, the need for private speech decreases. After they begin school, children gradually give up talking aloud to help themselves as they master new techniques such as writing or solving arithmetic problems.

Flexibility

To err is human, the old saying goes. Being able to set up the best strategies won't help if you aren't able to correct mistakes, to switch lanes and try a new approach. Bruce Hood and his colleagues investigated children's ability to let go of an expectation when it turned out to be wrong and follow a new line. The investigators prepared an apparatus consisting of an open rectangular frame with three chimneys on top and three boxes lined up below. An opaque tube went from one chimney diagonally to the box on the far side. When the experimenter dropped a ball down the chimney, children under three years old repeatedly looked for the ball in the box directly below the release point. They expected the ball to fall in a straight line. When the experimenter exchanged the tube for a transparent one and they were able to see the ball's descent, they found the ball in the correct box. However, when he again used the opaque tube, they were unable to generalize from their experience. They still looked directly under the chimney where they had last seen the ball. They thought it had to be there. Children over six years old were able to free themselves from this idea and mentally follow the direction of the ball through the opaque tube. They were able to keep an eye on both their own actions and the outcome and to change course if necessary.

Six-year-olds are generally capable of more flexibility than younger children. Just think back to what happened when you tried to change a detail in your toddler's bedtime story. Three-year-olds are notoriously inflexible when it comes to changing routine. This may be because they are just getting used to the fact that events take place in a particular order, and they want to hold on to that order. P. D. Zelazo, D. Frye, and T. Rapus asked children to sort a set of pictures either by color or by shape (category). After they had done it for a while successfully, the investigators told them to use a new rule. For example, if they had been sorting by color, they now had to sort by shape. The three-year-olds were able to state the new rule, but they continued to do their sorting by the old rule. The four- and five-year-olds were able to make the switch.

Learning from experience

I once knew a kind-hearted mother who tried to shield her child from any negative experience. Whenever little Stephanie forgot her gym shoes, her mother hired a taxi—no small expense for a woman of her circumstances—and delivered the shoes to her daughter's kindergarten. This and many similar experiences deprived Stephanie of the valuable opportunity to learn that her own actions, or nonactions, have consequences. How easy it would have been for Stephanie to learn this as a young child! She would have had to sit alone on a bench for half an hour while her classmates played and danced. She would have been unhappy for half an hour, but she would have experienced for herself the result of her own forgetfulness. Taking along her gym shoes would have been her own responsibility and remembering them would have contributed to her sense of her own competence.

Learning from experience that one's choices lead to a positive or negative outcome makes it more likely that a child will avoid harmful risk situations and invest more time in occupations that will support the attainment of long-range goals. Give your child a chance to learn this valuable lesson.

The Starring Role of the Prefrontal Cortex

During the burst of progress that takes place in a child's executive functions and learning abilities, the prefrontal cortex takes center stage. Much of what we know about the role of the prefrontal cortex comes from studies of patients with injuries to this part of the brain. Damage to the prefrontal cortex in childhood may lead to later problems in drawing conclusions (inferences) from available information or to other deficits in executive functions. An example is a little boy who was struck in the forehead by a lawn dart when he was four years old. Neuropsychological tests at five, six, and seven years all showed him to be of above average intelligence. However, he had severe problems in planning ahead, in solving novel problems, in making inferences, and in learning from experience. This indicates that the prefrontal cortex is involved in the network that is important for these functions.

Studies of patients with brain injuries show that the frontal lobes also participate in the emotional network. Lynn M. Grattan at the University of Maryland, Paul J. Eslinger at Pennsylvania State University, and Antonio R. Damasio, Daniel Tranel, and Hanna Damasio at the University of Iowa reported that adults who suffered lesions of the frontal lobe before the age of seven years not only had difficulty learning from experience and using executive functions but were also very impulsive and subject to emotional outbursts. They showed great fluctuations in mood, were less tolerant of frustration, and seemed to have a diminished sense of empathy. Their deficits made it difficult for them to keep their emotions under control and get along with other people, so they were apt to be outsiders.

The rapid development of the prefrontal cortex

The early childhood years are a time of intense construction in the prefrontal cortex. During a child's first three years the neurons in the prefrontal cortex grow a tremendous number of dendrites. At the same time, tiny projections called spines develop on the dendrites, providing sites for synapses with other neurons. The number of synapses increases rapidly, reaching a peak at around the age of three and a half years. This number is one and a half times as many as a child will have when he is grown up.

Important biochemical changes that have an effect on thinking are also taking place in the prefrontal cortex. There begin to appear special neurons that contain acetylcholine, a neurotransmitter involved in memory formation. The cells also appear in the associational cortices, areas in which sensory input is interpreted and put together with information from other areas of the brain. These cells increase in number until young adulthood and remain, thank goodness, until well into old age. The late and extended development of these neurons seems to be peculiar to human beings.

A good way to see the rise in activity in the prefrontal cortex is by measuring the amount of glucose that a brain region is consuming. The highest glucose consumption is observed in the frontal cortex between two and four years of age, indicating that a lot of construction is going on in that structure at this time.

The prefrontal cortex gets a boost from the rapid strengthening of the connections between the two hemispheres. The fastest growth in the part of the corpus callosum that links the frontal network of the left and right hemispheres takes place between the ages of three and six years.

Of blooming and pruning

A few weeks ago an aunt of mine called me long distance from Canada with an urgent question. Her daughter-in-law was planning to send her three-year-old child to nursery school, and she desperately wanted to be sure not to miss "the most important years for stimulating the child's cognitive development." She was afraid that windows of opportunity would close.

We sometimes get carried away by the excitement of hearing about the tremendous increase in synapses, the blooming that goes on during a child's first three and a half years, when maximum density is reached in the prefrontal cortex. It is easy to overlook the extremely important structural organization that takes place during the following years. During the pruning phase, which lasts until about age 19, synapses are eliminated faster than new ones are formed. This phase is an enormously fertile time for learning. In contrast to the blooming period, it is during pruning that the training of new skills takes on special importance. The structural changes in the brain depend very much on experience (experience dependent). Brain development provides the foundations for increasingly complex mental functions. And these new competences in turn foster brain development.

Memory Has Different Forms

Emily's teacher was impressed by how well Emily could remember the names of plants and the conditions that make plants grow. On weekends, Emily had been helping her father in the garden plot next to their garage. While she passed him his tools or dug little holes for the seeds with her trowel, she asked him all kinds of questions. It was surprising how much information she absorbed this way. Emily's knowledge of plants and gardening involves what is called "explicit" memory. This can refer either to words and facts (semantic memory) or to events (episodic memory). As language skills develop, a child's store of explicit memories grows.

While Emily was keeping her father on his toes with all her questions, her brain was busy storing all the new information he was providing. When she heard the name of a new flower, the information entered her memory's short-term storage facility. Her hippocampus helped her keep the memory for hours or days until it was transferred to long-term storage in her cerebral cortex.

When Emily sees the same flower again, her working memory retrieves the image of the flower she saw before and keeps it online while she compares it to the flower she sees in front of her now. Improvements in working memory may be responsible for the fact that six-year-olds are better able than younger children to keep two categories or different thoughts in mind at once and use them for weighing new information.

Working memory is fantastic all by itself, but what's in the existing memory store counts as well. Experiments have shown that children are better able to solve problems when they can use knowledge gained from previous experience, when they are already familiar with items or methods used. As a child grows, his brain can process information faster, and at the same time, his store of memories expands.

The hippocampus, which is a principle participant in the formation of memories, undergoes rapid changes during the toddler years. The neurons in the hippocampus develop highly complex dendrites, specialized for a particular kind of electrical activity involved in memory formation. In addition, the connections between the hippocampus, cortex, and limbic system are rapidly myelinating between the ages of two and three. This linking of areas important for knowing and feeling is probably why emotions act as a "glue" that makes memories "stick." Emily enjoys working with her father in the garden, so she likes to learn the names of the plants. Of course, a negative experience can also act as a glue. Emily won't forget the time she fell off her bicycle because she turned around too fast.

If we were to concentrate only on what children actively

remember and report, we would be neglecting the vast amount of learning that takes place without any conscious awareness. Maybe you weren't paying much attention to the cereal advertisement on television, but when you get to the store, your arm automatically reaches out for the box that was shown on the television screen. Memories like this belong to "implicit memories," memories that do not have to be stored on purpose or consciously recalled. Implicit memories have a powerful effect on habit formation. They most likely greatly outnumber the explicit memories, but both types of memory act together to influence a child's behavior.

Paul M. Fischer and his colleagues in Georgia and North Carolina reported on a study of implicit memory with practical implications. The investigators showed children brand logos collected from a variety of printed sources. These logos represented both products targeted to children and products primarily aimed at adults, including two cigarette brands. The children were asked to match cards with a logo to a picture showing the product advertised. It was important that none of the logo cards actually showed the product itself; for example, the Marlboro man was not smoking. One-third of the three-year-olds correctly identified Joe Camel with cigarettes, and in the six-year-olds the number of correct responses for Joe Camel was up to a whopping 90 percent. The investigators were not surprised that the children recognized the McDonald's arches and the Chevrolet and Ford logos. Automobiles are frequently advertised on television, and the children had most certainly eaten hamburgers at McDonald's. However, cigarette advertising no longer appears on television. Yet, by six years old, the children recognized Joe Camel as well as they did Mickey Mouse.

This experiment illustrates the power of implicit memory. The children had not specifically learned the common brand logos; they had picked up the information unconsciously, presumably from their families, caregivers, or the friends of one or the other. In much the same way, they take over habits and attitudes from their role models.

Memories of childhood

Try to remember your very earliest childhood experience, one that only you could possibly remember because you have never seen it in a photo, nor has anyone ever talked to you about it. Chances are that it won't be before you were two years old: it is most likely from the time you were about three to four years old, and it is possibly only from your early school years. We saw that six-month-old babies could remember a mobile trick for 14 days, so they can form memories. Why is it that although very young children can form memories, so few of these can be recalled in adulthood?

Many ingredients go into forming the kind of memories of past events that we can consciously recall years later. Important development takes place in early childhood. As a child grows, he becomes able to store more memories, and he accumulates more experience. The emergence of a sense of self-awareness makes it possible for him to link events to his own person. His growing ability to make more complex inferences helps him form more associations and apply these to new situations, thereby increasing the range of cues that can serve as reminders.

Being able to use language gives a child a means of categorizing and storing impressions in a form that makes them easier to call to mind voluntarily. Studies have shown that if a child is able to describe events at the time they occur, he is more likely to be able to recall them later. If he can discuss his experiences with adults, he can organize his memories better in a way that helps him to retrieve them.

The collection of mental abilities leading to the appearance of autobiographic memory rises on the foundations built up in the nervous system. During the second year, self-awareness, inference, and language unfold at a time when the two hemispheres are in an intense phase of being linked together.

According to Charles A. Nelson, at the Institute of Child Development at the University of Minnesota, one of the reasons infants are unable to recall their memories from storage could be because areas of the frontal and temporal cortex do not reach the necessary degree of maturation until around four years of age. By that time, a network of fine connections links the cortical areas within each hemisphere more tightly together.

Autobiographic memory is more to children than just a collection of stories in an album. Being able to remember events they experienced in the past is a basis for forming ideas about the world and about themselves.

Imagination

As children expand their language abilities, develop more complex ways of thinking, and collect more and more memories, they show an incredible spurt in their powers of imagination. Six-year-olds Steven and Sonja have decided to make an expedition to Mars. Steven sets up a couple of wooden stools to serve as rockets. Sonja gets some nuts and raisins out of the kitchen to take with them for food. They both put on their bicycle helmets and get ready for takeoff. They begin the countdown and make the rocket noises. In a few seconds, they have traveled through space and can descend to the surface of the new planet. They set out to explore, taking their "space food" with them. Around the corner, they encounter strange creatures that seem friendly and that even speak English.

Steven and Sonja are able to transcend the limitations of their surroundings. Their world expands. One of the ways they do this is by using symbols. They use one object for another. The stool becomes a rocket. They discover that words are also used as symbols. Steven can tell Sonja that they are now headed out through outer space and that he sees the red planet in the distance. She knows what he means because she knows what his words stand for. The two children use words to build a whole imaginary scene and to skip over the boundaries of time and space.

The children have come a long way since the time when 14-month-old Steven pushed his block of wood along the floor to imitate a car and 18-month-old Sonja fed her teddy bear and put him to bed. As three-year-olds, they pretended they were grown-ups preparing and serving a dinner. Now, as six-year-olds, Steven and Sonja are using their imagination to perform a sequence of acts they have not seen in daily life. They are able to create their own story, not of things as they are but as they might be.

Seeking an answer

Just as powerful as a preschooler's desire to master a task is his drive to find out more about his world. He wants to uncover the reasons behind everything. The time roughly between the ages of four and five is the "age of why?" Some children never cease to challenge their parents with inquiries ranging from "Why can't I have dessert first?" to questions on the great mysteries of life that have puzzled philosophers for centuries.

Here is the example of a five-and-a-half-year-old boy—from a home where natural science was a frequent topic—who had heard about God in his kindergarten.

> Daniel: *Who is God? Is he the one way up in the sky who watches to see if anyone is naughty? Could we see him if we went way up high in a big rocket? Is he very small? Could we see him with a microscope?*
>
> Mother: *No, he is a spirit, all around us, and you can't see him.*
>
> Daniel: *Is God a gas?*

Parents can encourage why questions and give answers appropriate for the child's ability to understand them. A four-year-old needs a simple explanation. If he asks why the moon gives light, just tell him it's because the sun is shining on it. A six-year-old can sometimes be encouraged to think of answers himself. If he asks why he has to learn arithmetic, you might ask him to think of situations where it could be useful. Parents can suggest looking at a question from a different perspective or another person's point of view.

The Cortical Areas Communicate

Underlying the growing complexity of a child's thought processes is the increasing ability of the different areas of the cerebral cortex to communicate with each other. One of the reasons communication improves is because the long cortical association fibers are myelinating, making them more efficient. Another reason is that electrical activity among different areas is becoming better synchronized.

Modern electroencephalographic (EEG) techniques show how electrical synchronization becomes more precise as a child grows. When we talked about how a baby learns to see, we mentioned the special high-frequency electric patterns known as gamma waves. These may be important for putting visual images together, a process that involves the whole visual cortex. The components of an image are registered in up to 32 areas of the visual association cortex. The brain has to bind them together into a picture. The most frequently discussed theory today of how it does this is called *coherence*. According to this theory, groups of neurons representing the separate components of the picture fire at the same time at the same frequency. This "chorus" of neuron groups binds the elements of the image together to allow the perception of the picture. The last step, how the neural activity leads to an actual image, is still a mystery. It is interesting that the high-frequency gamma waves seen when people view a picture can also be observed when they are dreaming.

We can apply the idea of coherence not only to putting visual images together but also to forming associations between bits of information from the various specialized areas of the cortex. Groups of neurons in different cortical areas fire synchronously at the same frequency. This may mean the visual area and the motor area if a child is about to reach for a toy, or the auditory area and the frontal cortex if the child is remembering a melody he heard several weeks ago. Activity that causes the groups of neurons to fire together not only makes the synapses between the cells within each group stronger; it also makes the long-range communication among the groups in different cortical areas more efficient.

An increase in coherence reflects the growing collaboration of different cortical areas and is probably a neural basis for learning by making new associations. EEG measurements in children show that coherence between the frontal and posterior lobes increases rapidly during the preschool years and first years of elementary school. Children with Down's syndrome do not show the strong spurts in coherence, indicating that cortical connectivity is affected in these children.

The Big "M": Motivation

We can't conclude our excursion into the paths of preschool thought and action without considering the forces that spur your child to solve problems or master new skills. We can think of motivation—which is related to the Latin word meaning "to move"—as the "motor" that boosts a child on his way to mastery and discovery. Toddlers already demonstrate a desire to perform tasks on their own and to achieve some degree of competence in doing them.

A few weeks after Emily's birthday party, the mothers meet for coffee at a local coffee shop. When Sonja's mother proudly remarks that she never has to remind Sonja to practice her violin, Steven's mother groans. Steven fussed for months before they gave him the drums for his birthday, and since then they have just been lying in the corner collecting dust. His mother sighs when she thinks of the energy with which he begins a project and the speed with which his enthusiasm evaporates. While Sonja shows exceptional perseverance, perhaps because of the particular skill that comes naturally to her, her example illustrates most clearly the development of a sense of competence in children about to enter school.

Preschoolers are on their way to becoming their own critics. They evaluate their current performance in comparison to their performance in the past and enjoy mastering the challenge of improving their skills. Monitoring their own progress builds a sense of competence and gives them the encouragement to take on new tasks.

Sonja also derives a strong sense of pleasure from her activity itself. While she is playing, she concentrates on performing the correct movements of her fingers, on listening to the tone. The intense concentration children exhibit when they are performing activities that interest them is striking: they are completely immersed in what they are doing, often oblivious to all that is going on around them.

The American psychologist, Mihaly Csikszentmihalyi, who has investigated the phenomenon of creativity in adults, has called this state *flow*. If children are able to immerse themselves in the intense pleasure of performing an activity for its own sake, the enjoyment they experience will encourage them to pursue such activities in future. They will be less likely later to turn to pursuits that are harmful or wasteful.

Very important for motivation is the close connectivity among cognitive, emotional, and executive systems. The areas that register pleasurable sensations are thus in an optimal position to affect the desire to perform an activity.

To Think About

Time for music lessons?

A great deal of discussion centers on the question of whether early music training has an effect on the acquisition of other skills. It is such an intriguing question that it will undoubtably be pursued over the coming years using behavioral studies combined with modern imaging techniques. For the moment, scientific evidence is not clear enough to support specific practical recommendations. However, our personal opinion supports music programs based on the child's current stage of development. Through participation in a music program, a child trains his sense of hearing, develops a sense of rhythm, exercises his memory, learns to concentrate on a task, and practices working with others. Most important is the joy he experiences in producing or listening to music. Music is such a basic part of human life that it should be enjoyed in its own right and not just as a means to an end.

The decision to begin instrumental music instruction depends on the child's own desire to play an instrument and the satisfaction he has in improving his skills. Training can improve hand–eye coordination and control of hand and finger movements. However, it is important to consider the child's current stage of development with respect to fine-motor abilities in deciding on which instrument to play and when to begin instruction.

Should I do something about handedness?

It is often difficult to tell whether your child is left- or right-handed. Children may use one hand to feed themselves and another to hold a pencil or to throw a ball. If your child shows a consistent preference for the left hand, help him find the best position for him to write. Since studies have shown a modest correlation between left-handedness and increased risk of accidents—perhaps partly resulting from the fact that the world is set up for right-handers—help your left-handed child use knife and scissors (special ones for left-handers) and put special emphasis on safety procedures.

Learning for school?

Attitudes about the right age at which to begin early systematic training have changed through the centuries. The Puritans thought it was essential to teach two-year-olds to read the Bible to ensure the salvation of their souls in the event of early death. Prominent educators of the nineteenth century took issue with this, suggesting that early reading leads to brain damage and insanity.

Although it is possible to teach five-year-olds—or even younger children—to read and spell, this kind of rote learning does not encourage active problem solving or the development of concepts. The desire to learn to read should come from the child. Observe what your child enjoys doing on his own. A special interest in a topic is a powerful motivation for trying to find out more about it. If your child is interested in dinosaurs, give him a book with good pictures of dinosaurs and a small amount of text. You could also take him to the library and let him choose a book.

More important than rote learning is the set of basic attitudes a child develops toward learning. Your example is contagious. Do you enjoy reading, drawing, listening to music? Do you want to know "why"?

Children enjoy mastering the challenge of tasks that demand a certain amount of concentration and effort on their part. You can sustain this enjoyment by adjusting the hurdles to the level that is suitable for your child's stage of mental development. It is better to play a game a child *can* win—for example, games of chance like Chutes and Ladders or Parcheesi or simple games of concentration, like Memory—than to play more difficult games and bend the rules to *let* him win. A sense of his own competence will give him the confidence to approach new tasks.

Should medication be considered for a hyperactive child?

Some children have extreme difficulty in sitting still and paying attention. A careful clinical examination is necessary before the diagnosis of ADHD can be made. Not every inattentive or restless child has ADHD. It is estimated that the disorder affects 3 to 10 percent of all school children.

With children clearly diagnosed with ADHD, initial treatment is psychological; it involves providing a calm, consistent surrounding and using behavioral techniques to help the child learn to concentrate. If the psychological approach is not sufficient, medication may be necessary. It is important for parents to keep an open mind on this question. A child may suffer because he is frustrated at his lack of success in school or embarrassed by the ridicule of his classmates. Carefully dosed medication may give him the chance to develop his skills and build his self-confidence. However, it is critical that a doctor monitors the dose carefully and regularly reevaluates the situation of the individual child.

How can I stimulate my child's motivation?

What can parents do with a boy like Steven who shows a lack of perseverance? For a start, limit distractions. Observe what kind of things he likes to do. Spend more time with him. Show interest in his activities, for example, by asking him to show *you* how the march sounds on the drum. It is important to give feedback that focuses on his effort rather than the quality of the final accomplishment. However, praise should be sincere. If lavished too generously, it may become useless or even harmful, either because a child feels it was not deserved or because he comes to depend on it.

Laptops for tots?

Whether or not we are personally attracted to them, computers are a fact of life in today's world. On the positive side, computers are the tools of modern life, and becoming familiar with them is almost like getting used to books and pencils. Computers have a big advantage in teaching skills on an individual basis. Children who have trouble learning the alphabet or basic arithmetic procedures can practice at their own speed and be instantly rewarded for their success. Computers have endless patience, and are not moody. They are fair.

Some games involve perception tasks (finding differences), memory (where is the treasure?), focused attention, reaction speed, hand–eye coordination, or impulse control. If two or more children are playing together, the computer may draw them into a discussion, encouraging them to use language to discuss strategies and solve problems together.

The most obvious point on the negative side is that while children are playing with the computer there are a lot of other important things that they are *not* doing. They are not reading books, drawing pictures on paper, playing the piano, or exercising their muscles by running, jumping, playing outdoors. They need opportunities to develop language and social skills by playing with others and to exercise their imagination. A question that is being explored is whether computer games stimulate only brain regions that are involved in vision and movement and not those that are important for learning, memory, and emotions.

Young children's visual systems are still undergoing development, and they need to practice all kinds of hand–eye coordination, not just pressing buttons. Jane Healy, author of *Failure to Connect: How Computers Affect Our Children's Minds for Better or for Worse*, mentions that children may not want to draw with markers and crayons because their pictures don't look as good to them as those on the computer.

Parents should monitor the content of the programs or games the children are using. What are the messages children are being exposed to? Many computer games present violence as an easy way to solve problems. Speed is more important than thinking about the consequences of an action. Brutality is acceptable because death and injuries are shown with no compassion for the victims. The reason behind the success of these games is that they are the ones that sell. As Kenji Eno, of the videogames division at the Sony Corporation, put it, "fear sells better than love." As a society, we should ask ourselves why. As parents we should explain to children (and to the entertainment industry) why we think a particular product is unacceptable.

How can I help my child's ability to solve problems?
You can occasionally give your child a boost by providing conditions that help him learn—without actually solving the problem for him. This is what is sometimes called *scaffolding*. The term means carefully changing the demands of the task so that your child has a challenge to overcome and gradually withdrawing your assistance as your child's independent problem-solving ability increases. Let's suppose your four-year-old brings his toy car to you because it doesn't work. If he doesn't tell you the reason, ask him what could be the matter. It could be the batteries. Let him change them while you watch. He may already be better at opening battery compartments than you are, but if he isn't, let him try first before you suggest something like, "Maybe you have to push and slide the cover at the same time." Instead of saying, "The batteries go in this way," let him put them in and see if the car goes. If it doesn't, ask what he could try now. Your interest encourages his persistence, and in the end he should be the one to solve the problem. The next time the batteries go dead, just give him the new ones and let him do the rest.

Living Together

"Let's go and see *everybody*," said Pooh. "Because when you've been walking in the wind for miles, and you suddenly go into somebody's house, and he says, 'Hallo, Pooh, you're just in time for a little smackerel of something', and you are, then it's what I call a Friendly Day."

A.A. Milne, *The House at Pooh Corner*

Pooh's spontaneous, heartfelt suggestion illustrates the powerful attraction of togetherness. As children grow from toddlers to schoolchildren, they experience the irresistible, warm feeling of participating in life with others. The desire to belong to a group shapes their social behavior. They eagerly absorb the language around them and use it to make new friends. Since their circles are expanding to include more people outside the family, they learn to adapt their ideas of rules for living together to a wider group. At the same time, they are gradually learning to guide their own behavior with less parental control.

198

Belonging to a Group

While toddlers were just becoming aware of themselves as persons, preschoolers are becoming aware of themselves as members of a variety of different groups. They even begin to distinguish themselves from those outside the group by wearing special items of clothing, by participating in particular activities, or inventing a secret "language."

This is the time of budding friendships. Closer bonds are possible because children are not only becoming better able to understand the feelings and intents of other people but can also use language to talk about them. Having a friend gives a child a chance to experience the support of someone of her own age group, gives her a chance to learn what it means to keep a friend. Children's concepts of loyalty and mutual obligation are expanded from the close family circle to their new groups.

Around the ages of five to seven children begin to compare their playmates on the basis of strength, size, looks, skills, and, well, yes, possessions. While children develop an initial set of standards in their interactions with parents and older children in their families, they are now confronted with the standards of a peer group. Belonging becomes very important. They are acutely sensitive to the opinions of their peers.

It is often very tempting to say things like, "Steven was always so cheerful and helpful until he started playing with Tommy more often." Children do indeed learn from each other and often imitate the behavior of a peer. However, it is important to remember that they also actively *choose* their companions. Steven needs his parents' help to see why he shouldn't automatically imitate everything Tommy does.

A child learns through her experience in daily life that belonging to a larger group involves many of the same things that she experiences in relationships with family members and with friends: taking turns, sharing with others, helping to get a job done, and comforting or aiding others in times of distress. It is not enough just to be physically present. It is not enough merely to perform one's tasks. The greatest satisfaction is derived from a sense of being needed. Having the opportunity to experience the intrinsic rewards of participating as a valued member of the group is a powerful factor in building a child's sense of self-esteem.

Some of a child's behavior is clearly based on self-interest. Taking turns means you will get your turn when the time comes. A child might also give a friend a present, hoping to play with it himself or with the vague idea of receiving a similar gift in the future. This is the give-and-take of daily life and is not unfamiliar to adults. However, children are also capable of altruistic behavior, of doing something that has no apparent advantage for themselves. This can be comforting an injured friend, picking a bunch of flowers for a mother with the flu, jumping up to find a father's misplaced keys. They act out of a spontaneous impulse to make someone happy. A prerequisite for this response is their ability to sense what others feel, their capacity for empathy.

In addition to empathy, children feel actual "concern." They want to do something to alleviate a situation. This means putting themselves, at least temporarily, in the background and focusing on the other person, going from an egocentric point of view to one that includes the welfare of a wider group. The psychologist David Hamburg claims that caring social behavior learned in childhood opens the way to constructive human relationships throughout the life span.

Caring social behavior is not something that necessarily develops all by itself. A friend's story about his teenage son showed me better than any theory how an egocentric approach can thrive if parents do not offer any guidelines. The boy had an after-school job in a grocery store to earn spending money. All his expenses were paid by his parents, and they were putting aside part of their modest salary for his college education. The boy liked to eat well, so he occasionally used his spending money to buy himself a large, juicy steak, something the parents could not afford. After the boy's mother fried the steak for him, he ate it in front of the family without offering any to them. He took it for granted that since he had paid for it, he was entitled to eat it by himself. I was surprised that the parents didn't speak up and tell him what being a member of a family means. Sharing and caring need to be practiced and encouraged at home as well as in kindergarten and school.

Talking and listening

A child's growing language abilities add new dimensions to life in the group and support the building of closer personal relationships. By six years of age a child's speech can be understood by everybody around her. She shows that she has an idea of past, present, and future by telling stories about events that happened and talking about what she will do tomorrow. This ability helps her remember events and to learn from experience. Children can use language for a multitude of purposes, whether it is to inform (I have to go to the bathroom *now!*), to request (Can I have another piece of cake, please?), to get attention (Mommy, watch this!), or to share a joke.

Six-year-olds are capable of talking about emotions and intentions and about the connection between cause and effect. They can say, "Jimmy is happy because his father is coming home tomorrow," or "Susie is crying because her doll is broken." A child's use of language is now sufficiently precise to allow her to talk about abstract concepts such as beauty, stupidity, or fairness. They can discuss simple moral dilemmas.

Communication is a two-way street. Using language means listening as well as speaking. Children have to learn not only to express their own thoughts and feelings more precisely but also to interpret and respond adequately to what others are saying. Giving children a chance to practice this conversational back-and-forth and to learn from the way adults listen and respond to each other improves their ability to use language in a variety of social situations. The best way to improve children's language skills in today's busy world is to take time to sit together at meals.

Fairness

"Tommy hit me!" "Emily stole my felt pen." "Steven has a bigger piece of cake." "It's not fair!" Children seem to talk as much about justice as the Supreme Court. It's not only repeating words they have heard adults use. Preschoolers are finding out for themselves that living in a group entails guidelines for getting along with others.

One of the ways we see this is in their behavior in competitive games. They have a strong sense of rules, emphasizing that cheating or changing the rules without the assent of the group is "unfair." If one of them does not comply with the accepted set of rules, the group may refuse to let her participate.

At the same time they are learning about rules, children are also learning that living together means taking more than just rules into account. They are beginning to develop a moral sense, a feeling that certain actions are intrinsically right or wrong, independently of whether adults are observing them. They also begin to take both rights and needs into their consideration of moral dilemmas.

Kimberly Wright Cassidy, June Y. Chu, and Katherine K. Dahlsgaard, at Bryn Mawr, presented preschool children with a series of situations in which they were asked to make a "moral" decision. One example was the story about two boys who were sitting in a doctor's waiting room. The first boy had been waiting patiently for his turn to see the doctor for a long time. Just as the nurse was about to call him, another boy arrived with a cut on his arm. He was groaning with pain, and tears were streaming down his face.

When the investigators asked the children who heard the story which boy should now see the doctor, they mentioned that the first boy had a right to go in first because he had been waiting longer. But they felt the second boy should see the doctor because his need was greater. Children are able to see both the justice and the caring point of view. They are able to understand the principle of rules and, at the same time, the needs of the boy who was hurt. The researchers found no differences between boys and girls.

The fact that the children who heard the story about the two boys in the doctor's office were willing to accept that one person's needs can take priority over the other person's rights shows that their sense of right and wrong is based on empathy, on the ability to feel for the person in need.

From the age of about three years, children develop a real awareness of the feelings of other people. This is seen in the fact that they show signs of guilt not only when they have injured another person physically or damaged someone's possessions but also when they have said something that caused the other person to be sad.

As a child approaches school age, her moral sense grows to include the realization that she could step in and prevent harm to another person. A child might see a bully beating up a weaker child. She knows the bully's actions are wrong and that the other child needs her help, but she is afraid to step in because the bully is so much stronger. The child may feel guilty that she didn't help.

The fact that children are aware of moral standards does not mean that they will necessarily always act according to them. However, they should be able to understand—know *and* feel— that an action was wrong. A friend of mine told me of an incident that impressed him when he was a young boy with the habit of snatching his younger brother's toys away from him. When his father caught him in the midst of yet another blatant attempt at "might makes right," he set my friend firmly down on a stool, looked him straight in the eye, and said "Think!" With that one word his father had expressed his faith in his son's ability to turn his behavior around on his own. It was a lesson the son never forgot.

Overcoming prejudice

A witch is bad. What would you answer—yes, no, or maybe? Very young children would instantly answer "yes." Being able to think in such stereotypes would certainly have been helpful in evolution. You wouldn't have had time to think about whether the wild bear was going to eat you or wonder about the intentions of the stranger approaching you with his club. For babies and toddlers, simple categories are a help in beginning to make sense of all the new impressions in their world. However, this restricted manner of thinking becomes more and more of a hindrance as children move out into the complex world of other individuals.

An ingenious experiment making use of a fairy-tale situation showed that during the preschool years, children become progressively better able to move away from rigid, simple stereotypes. Cynthia Hoffner and Joanne Cantor of the University of Wisconsin–Madison prepared four pictures. The first showed a pleasingly plump stereotype of a kind grandmother lovingly cuddling a cat in her arms. The next showed the same grandmother angrily holding the cat up by the scruff of its neck. The third picture was of a thin, ugly woman with a pointed chin and crooked nose, the common stereotype of a witch. The "witch" had a kind expression and was holding the cat affectionately. The last picture showed the witch angrily holding the unfortunate cat by the scruff of its neck.

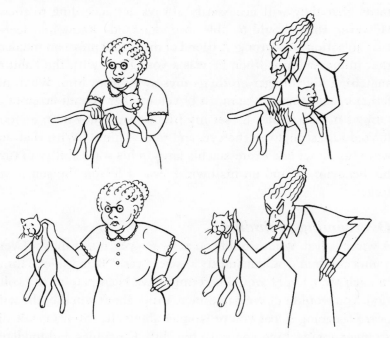

Stereotype Test: Judging a person as "nice" or "mean" based on her appearance or based on her actions.

The investigators then showed the pictures one at a time to the children, asking them to say which woman might be likely to invite them in to have some cookies. Would the children like to go in and visit her? In spite of what they saw the grandmother do to the cat, the four-year-olds were sure the grandmother would be "good" and invite them to stay and eat cookies. No matter what the witch did, she was always "bad," and the children didn't want to stay with her. They were judging on the basis of appearance only. The six-year-olds, however, were more likely to consider both the woman's appearance and her behavior, and to judge the behavior, not the appearance. The flexibility shown by the six-year-olds represents a big step in a child's cognitive abilities and is a basis for overcoming prejudice.

Just because children have the mental capacity to overcome a fairy-tale stereotype when presented with contradictory evidence doesn't mean they will automatically do so in real life. This is because stereotypes most often represent an accumulation of associations that have built up implicitly, or unconsciously, over a longer period of time. For this reason, stereotypes can be very stable. Both reasoning and experience are necessary to break them down. Of course, overcoming prejudice is more than a matter of correcting a child's stereotyping; it is a lifelong quest. This is especially crucial as our world expands to become a global multicultural society.

Taking Over the Controls

Tommy's kindergarten teacher feels discouraged. Tommy is rowdy and a constant troublemaker. If she as much as hints that he should wait to take his turn, he becomes angry and abusive, both to her and to the other children. Yet several times his questions about animals have shown that he is not only very observant, he also has a good memory, and he often surprises her by coming to some very discerning conclusions. Once he finally decides to sit down and draw a picture, it often turns out to be dynamic and imaginative. She finds him intelligent and perhaps even talented.

But with her years of experience, Tommy's teacher worries that the boy is not his own best friend.

An important part of growing up and getting along with other people is acquiring the ability to find one's way in the thicket of emotions and impulses that are a part of our lives. Children who learn to overcome frustration, keep the upper hand over impulses, curb their aggressions, and develop the self-confidence to take on problems as challenges start out with a great advantage.

Overcoming frustration

Ever since Sigmund Freud presented his pioneering theories on the influence of the unconscious on human behavior there have been tendencies to equate frustration with something bad, to assume that frustrations lead automatically to neurotic behavior later in life. The conclusion was that situations that might lead to frustration should be avoided at all costs. However, this isn't quite what Freud himself said.

His daughter, Anna, clarified Freud's position in her series of lectures held at Harvard University in 1952. She emphasized the crucial role of parents and other educators in helping children to make an extremely important step in their development. Between the ages of three and five years, children should learn to stop and think before reacting to their immediate desires. She went so far as to say that a child develops inner strength by overcoming frustrations rather than by avoiding them. This was not just part of a "keep-a-stiff-upper-lip" philosophy: because frustrations belong to human life, children should be allowed to learn how to cope with them.

Five-year-old Steven began to cry when Emily told him that his drawing of a horse looked like a pig. He had been trying so hard, yet he felt his efforts had failed. He was frustrated and ready to give up. He started to tear up his paper. Many such experiences over a long period of time could discourage Steven from trying anymore. Luckily, Emily's mother was there. She didn't draw Steven's horse for him. Instead, she showed Steven how he had started his drawing wrong. He had just dashed off with a circular motion without stopping to think about the horse's shape. She let him see that he would need to start with a rectangular form. This "hint" was what Steven needed to get him off to a better start, and in the end he was also quite pleased with his horse. If Steven has many similar experiences, he will strengthen his persistence and develop enough confidence to try new solutions because he has seen that these can be successful.

Controlling impulses

Being able to control an impulse gives a child the opportunity to stop and reflect long enough to avoid a poor strategy and employ one that might be more successful. Impulse control is a key to flexibility in thought and action. The importance of impulse control is not limited to short-term problem solving. It can also be extended to the development of strategies for postponing an instant reward in the interest of reaching a goal sometime in the future.

Yuichi Shoda, Walter Mischel, and Philip K. Peake carried out a series of illuminating experiments in which four-and-a-half-year-old children could decide how to deal with the problem of postponing a temptation. The child was seated alone in a small room at a table with a bell on it. The examiner then placed something the child was especially fond of—for example, a marshmallow—on the table and said, "If you wait until I come back without eating the marshmallow, I'll give you two of them. You can ring the bell at any time, and I'll come back; but then you won't get the marshmallow but a pretzel (something the child didn't care about) instead." If the child didn't ring the bell, the experimenter returned after 15 to 20 minutes and gave her the promised reward—if the marshmallow was still there.

For some children the temptation was too great to resist. They ate the marshmallow. Others, however, held out until the investigator returned. They used little "strategies" to help them reach their goal: closing their eyes, looking away, talking to themselves, or playing.

The authors of the study did not come to the grim conclusion that the important message of the marshmallow test is the ability to forgo pleasure under all circumstances. Instead, they suggested that a child must have the freedom to choose and not be automatically compelled to follow an impulse. Some children have a particularly hard time waiting for something they want badly. This depends in part on the child's temperament. However, children are capable of learning through experience to tolerate increasingly longer delays.

Impulsivity and later delinquent behavior

Impulse control merits special attention because of correlations between high impulsivity and later delinquent behavior. Richard E. Tremblay and his colleagues at the University of Montreal found that boys who showed a particular cluster of behavioral characteristics in kindergarten were more likely to show delinquent behavior as young adolescents. The three characteristics were high impulsivity, low anxiety, and low reward dependence. *Reward dependence* is difficult to define because we automatically think of tangible rewards. Here, however, it means a child's sense of warm, social ties with caregivers and a need for their signs of encouragement or approval.

Boys who showed the same first two characteristics but who also showed a high degree of reward dependence were less likely to turn to delinquent behavior as adolescents. It is possible that their relationships to their parents or other caregivers were a decisive factor in shaping their behavior. The authors recommend paying greater attention to preschool children with at-risk profiles.

Curbing aggression

Reports of an increase in violence and aggressive behavior among schoolchildren are of great concern to parents. It is not so much the "rough-and-tumble" energy that tends to surface in young children, particularly in boys, but rather behavior that is clearly—often even intentionally—dangerous to others.

Aggression has a sort of natural history. The tendency to use physical force to influence the outcome of a situation builds up from the age of nine months to two years. It is a comfort to know that aggressive behavior generally decreases from ages three to five, followed by another drop between 16 and 20. The fact that some children are more prone to aggressive behavior than others is discussed in Chapter 10.

During the preschool years children normally learn from adults what behavior is acceptable and what is not. They are also encouraged to express their needs through language and made aware of the feelings of others. If this guidance is missing, or worse yet, if the adults in their lives actually model violence or brutality, they will be more likely to incorporate these patterns into their own lives.

Developing self-confidence

I was very impressed by a four-and-a-half-year-old boy who participated in an experiment carried out by the Harvard Infant Study to investigate the development of temperament. The experimenter showed the boy a picture and said something like, "This is my favorite picture. I really like it a lot." Then she told the boy to tear up the beautiful picture. While the other children complied with the investigator, either playfully or dutifully, this boy spoke up and said, "I don't want to do this." To him, tearing up the picture was the wrong thing to do, and he had the self-confidence to express his opinion.

Children need to feel sure of themselves with respect to their feelings of right and wrong in a variety of situations. They have to be able to say no to a stranger who invites them to come along. They may have to stand up to a playmate who is too rough with a weaker child. Self-confidence also means learning to take responsibility for their own behavior and having the courage to speak openly about their mistakes. These are strengths that do not necessarily develop by themselves. They need adult guidance and plenty of practice.

The Hemispheres Work Together

Nature has provided us with an ingenious solution to the problem of processing an enormous amount of information and handling the complexities of human behavior: two parts of the brain that are similar, each contributing special areas of expertise. This pooling of resources results in a huge network with far greater potential than the sum of the two separate hemispheres.

However, both hemispheres constantly work together. Trying to use one without the other would be like trying to march on one foot. The two hemispheres communicate with each other mainly over the corpus callosum, the strong bridge that links them, but other, smaller bridges exist as well.

Until now, much of what we know about the roles of the two

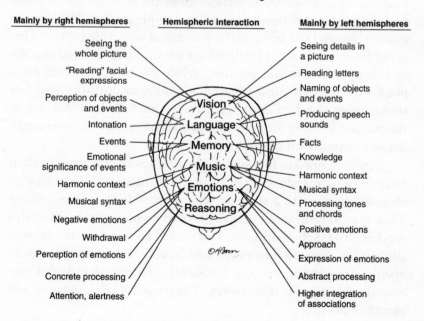

The Hemispheres Work Together

Mainly by right hemispheres	Hemispheric interaction	Mainly by left hemispheres
Seeing the whole picture	Vision	Seeing details in a picture
"Reading" facial expressions	Language	Reading letters
Perception of objects and events	Memory	Naming of objects and events
Intonation	Music	Producing speech sounds
Events	Emotions	Facts
Emotional significance of events	Reasoning	Knowledge
Harmonic context		Harmonic context
Musical syntax		Musical syntax
Negative emotions		Processing tones and chords
Withdrawal		Positive emotions
Perception of emotions		Approach
Concrete processing		Expression of emotions
Attention, alertness		Abstract processing
		Higher integration of associations

Nature's Ingenious Plan: The two hemispheres of the brain are similar, yet each contributes special areas of expertise.

hemispheres came from studies of brain injuries. Main disadvantages of this knowledge were that it was hard to pinpoint the exact location of the lesion, and by the time behavioral investigations were undertaken, compensatory changes in the brain had already taken place. Modern imaging techniques can now provide important insights on interhemispheric activity by observing normal activity in an intact nervous system. Evidence from these studies illuminates not only the roles of the left and right cortices but also those of subcortical structures, which are asymmetrical too.

Neither hemisphere is responsible by itself for a complex task such as language comprehension, memory, or emotional expression, to say nothing of global abilities summarized as intelligence or imagination. Instead, each hemisphere cooperates by contributing components that are part of the network necessary for the behavior. We can illustrate this with the simple analogy of a transistor radio. If you remove the battery, no sound will come out. The battery is a necessary component of the radio's functions: yet you cannot say that it is the battery that makes the music.

A complex behavior such as language can be broken down into many components. Speech sounds are registered and analyzed mainly by the left hemisphere. The language centers in the left hemisphere analyze the sequence of sounds and help access the words stored in memory. Meanwhile, the right hemisphere processes the melodic elements of the word flow and contributes emotional tones. It is interesting that language and music are processed in overlapping areas in the brain. New research shows that the interpretation of harmonic context as well as musical syntax activates areas in both the left and right hemisphere.

Electroencephalographic (EEG) studies showing brain activity in newborns given sugar water and in 10-month-old infants temporarily separated from their mothers reveal that positive emotions are mainly processed in the left hemisphere, while negative emotions activate mainly the right hemisphere. The right hemisphere is also more closely linked to reactions of the body to disturbing or painful situations. The left is likely to be more activated in children who are able to dampen their responses and to calm down quickly.

Differences in the relative amount of neurotransmitter activity help explain the variations in the roles of the left and right hemisphere. Norepinephrine, a neurotransmitter involved in alertness and attention, has a higher number of receptors and shows greater activity in the right hemisphere, as does serotonin, which plays a role in moods. Cortisol, a hormone released in stressful situations, also shows higher activity in the right hemisphere. Dopamine, a neurotransmitter crucial for working memory and for forming complex associations, has more receptors in the left hemisphere.

Changes in the relative amount of blood flow indicate a major step in a child's development. Until about the age of three years, blood flow is greater in the right hemisphere than in the left; after this time, it is greater in the left hemisphere. This may be an indication that the right hemisphere develops earlier than the left hemisphere and also reflects a shift in emphasis from behavior that is mainly driven by emotional reactions to behavior that is also influenced by the development going on in the language and higher order processing areas of the left hemisphere. The "thinking" areas of a child's brain step in and say, "Wait a minute." This gives her greater flexibility and a chance to make more use of her brain's resources. Her behavior can be described as reflective, rather than merely reflexive.

To Think About

What if she says, "They all do"?

Although you can't always predict what will come up next, you can inform yourself about what to expect. Sometimes the most enlightening experiences will come from watching the parents around you with their slightly older children. Talk with your partner about issues in advance and know how much leeway or room for discussion you are prepared to allow. What will you say when your child asks you for more pocket money, for guitar lessons, for her own television or phone? Don't be distracted by the words "everybody else's parents." Instead give your child a clear reason if your answer is no.

How can I boost my child's self-confidence?

We have only to look at our most popular fairy tales to see the attraction of themes that involve seemingly helpless individuals, often children, and their success in outwitting more potent beings like giants and witches or in otherwise overcoming what at first seem to be insurmountable obstacles. But we don't always have to look to folklore to find stories to inspire children to positive expectations about themselves. As psychologist Jerome Kagan has pointed out, stories about family members, past or present, give a child a strong sense of family identification. Knowledge of how other members of her own family have overcome difficulties can boost a child's confidence about her talent and chances for success.

How can I handle sibling conflicts?

Obviously you will have to step in whenever one child is at a serious disadvantage. Children are often not aware of their own strength or of the possible harm that could arise from the use of physical force to settle differences of opinion. However, in many cases the opposing parties are capable of finding good solutions if given a chance to think about them. For example, your six-year-old twin boys have been fighting about which one will have the top (or bottom) bunk bed. Ask them to come up with the solutions. They may suggest taking turns, or one may get the use of a disputed toy in return for giving in on the bed question. Giving *them* the opportunity to solve their conflict makes it easier for them to abide by their decision and gives them valuable practice for later.

In minor disputes of younger children (three- to five-year-olds) it's often impossible to distinguish the "aggressor" from the "victim." Here the no-fault policy saves wear and tear on your nerves and encourages children to resolve their disputes peacefully. Send both (all) parties to separate rooms and tell them to come out when they have found a solution and are ready to play together again.

What about medication for behavioral problems?

An article by Julie Magno Zito and her colleagues in the February 23, 2000, issue of the *Journal of the American Medical Association* (JAMA) called attention to the fact that prescriptions for psychotropic drugs for preschoolers are on the rise. Since these drugs affect the action of neurotransmitters at the synapse, we should be very cautious in employing these substances at such an active period of brain development. Joseph T. Coyle of the Harvard Department of Psychiatry has stated: "Given that there is no empirical evidence to support psychotropic drug treatment in very young children and that there are valid concerns that such treatment could have deleterious effects on the developing brain, the reasons for these troubling changes in practice need to be identified."

It is important to address directly the needs of the individual child, the child's parents, and those of the neighborhood, day care center, or school. Where all work together, much can be done without resorting to medication. However, a comprehensive assessment of the whole situation can determine that a child could benefit from medical help.

What options do I have if I feel I can't cope with my child at home?

If you feel you are at the end of your capacity to cope, it is better to seek the competent advice of friends or professionals with experience in helping families solve their problems. One promising method is that of video home training. Trained social workers make 8 to 10 visits to the family at home and document on videotape 10 to 20 minutes of daily life activities such as an evening meal or a family game. By pointing out the positive interactions between parent and child, they are able to build a parent's confidence in handling potential conflict situations. A father may suddenly realize that his son was actually looking at him full of admiration.

How can I make my child feel part of the team?

Three-year-olds like to think of themselves as part of the team. They enjoy helping in the house and are proud to have their efforts appreciated. They should be encouraged now; even if it does take longer to do a job than if you did it yourself. The team idea can be fostered in many ways. For example, if the whole family is going on an excursion, each member can carry part of the picnic, not just his or her own lunch. One person carries the lemonade, one the hotdogs, one the napkins and matches. Only when the whole team is together is the picnic complete.

As children get older they can take on real responsibilities around the house such as preparing meals, doing shopping, taking care of pets. Sharing in family rituals is another way of strengthening the bonds within a family. Not only children's birthdays but those of parents and grandparents, too, are a cause for celebration.

How do the two brain hemispheres collaborate in daily life?

Everyday life presents many ideal opportunities to observe and foster the unique interaction of the two hemispheres. Suppose you are looking with your child at a picture of a lot of trees around a lake. She will probably first see the picture as a whole (right hemisphere) and say, "It's a forest" (word: left hemisphere). You can encourage her to look longer and detect a surprising detail (left hemisphere), such as a little bird in a nest hidden in the branches.

Putting feelings into words is another occasion to exercise right–left cooperation. For example, if your daughter comes home sad because she didn't get an invitation to a friend's birthday party, you can help her label her feelings as sad, angry, or disappointed. Being able to talk about her feelings fosters her ability to understand the situation. Her sadness is primarily registered by her right hemisphere, but her left hemisphere contributes the appropriate words and helps her understand the reasons for a person's actions.

Memory is enhanced when both hemispheres are doing their part together. While the right hemisphere is more involved in memories for events as a whole, the left hemisphere is called into action for the details and for putting the memories into words. You can help your child remember more about an event by listening to her retell her experiences and helping her to remember details. For example, if the event was a visit to the zoo, ask her to tell you what the kangaroo was doing. You can also add a few interesting facts about kangaroos: They live in Australia. A kangaroo baby grows in its mother's pouch.

Paths to Personality

How might the children in our story have turned out later? We frankly are unable to predict. No crystal ball can transport us through the coming years and take into account all the factors that go into the growth of a child's personality. At times public opinion has shifted between attributing everything to genetics or everything to the environment; however, the picture that is emerging as a result of recent research in psychology, genetics, and neurobiology is a complex, multidimensional jigsaw puzzle of interacting factors. One way to sort them out is to divide personality into two main components, temperament and character. Temperament is that part of personality that has more to do with emotions and immediate reactions of the nervous system. Temperamental characteristics appear very early in life, and they remain relatively stable throughout the life span. Character, on the other hand, involves goals and values. Because these are influenced by the social and cultural environment and by a person's own experience, character takes shape over a long period of time.

Temperament

Remember how the two little girls, Anna and Emily, reacted to novelty at 4 months of age and again at 21 months. The Harvard Infant Study team paid particular attention to the development of children who, like Emily and Anna, were classified as belonging to the opposite ends of the scale representing reactions to the unfamiliar: the "Emilys," who were least disturbed by unfamiliar objects or events, and the "Annas," who were most disturbed by them. Over the subsequent few years, the investigators adjusted the experimental situations to the children's age by including play sessions with small groups of children, followed by parties with larger groups.

The children like Anna, who had been so upset by the strange voice and mobile at 4 months, thrashing her limbs and crying, and who were more fearful at 14 and 21 months, were more likely to be shy at preschool age than the children like Emily. The "Annas" were often more anxious and subdued. However, the investigations also found that these children made fewer errors on a task that required not reacting instantly to an impulse. They were more cautious and took time to size up a problem instead of rushing out with a hasty answer. The "Emilys" were more likely than the "Annas" to be outgoing and sociable preschoolers, more at ease and talkative with strangers.

When the investigators looked at the brain activity underlying temperament, they found that the brain patterns were remarkably stable as the years went by. Using electroencephalographic (EEG) techniques, they measured the electrical activity in the brainstem when a child heard a sound. At 10 years of age the children who had been described as high reactive at four months showed a higher reaction to the sound. The high response of the brain to sound was due to the greater sensitivity of the amygdala, which activates the brainstem. Their nervous systems still had a lower "threshold" for stimulation. However, the study held some surprises. Although the children like Anna were more likely to be shy than the "Emilys," most of the "Annas" were less shy than expected. They enjoyed playing with their friends and were happy in their school.

We can imagine that Anna never did become the life of the party, but as we leave her now, she is happy in her first grade class, admires her teacher, and has discovered the wonderful school library. As she grows, Anna's cerebral cortex, the "thinking" part of her brain, becomes a more powerful influence on her personality, allowing her to draw from her expanding store of experiences and to "jump over her shadow." History is full of examples of people who have overcome their extreme shyness and achieved fame as public speakers or on the stage.

Emily, who was a bouncy, sociable baby, an outgoing toddler, and a curious, enthusiastic preschooler, is looking forward to second grade. She never stops talking, and her conversation is punctuated by frequent laughter. We have seen her show empathy for her friend Anna, who was pushed off the tricycle, and we have seen her cheerfully helping her mother clean up after her party. Taking all factors into account, it is likely that her experiences in her new school will be positive.

Little Andrew, who led his parents a merry chase as a baby, is more persistent than his big sister Emily. But he has learned to take turns and respect the rights of others. He doesn't come home from preschool bubbling over with the day's events as she does, but he plays happily with his friends. His parents find it hard to believe that he was such a fussy baby.

Steven's mother has noticed that, unlike his big brother, with whom a reprimand just seemed to go "in one ear and out the other," Steven often displays signs of remorse. Once when she scolded the boys for breaking an antique vase because they were playing ball in the living room, Steven's brother just walked away, while Steven, feeling sorry for what they had done, suggested taking the vase to a shop to have it fixed or helping to pay for a new one. It is important to know that such differences do not depend only on parenting styles; they are the result of the unique interaction between the child's temperament and his experience. Research is currently exploring the reasons that children with a more anxious temperament are more susceptible to feelings of guilt at violations of standards. One hypothesis is that their nervous system predisposes them to physical feelings of discomfort. When they sense that they have done something wrong, they feel uncomfortable, or "bad." They feel the emotional reactions of their body more intensely—for example, a faster pulse, a dry throat, a hot feeling spreading over their face, a tenseness in the abdomen. This is consistent with the finding that children who reacted intensely to their inoculations at two months were also more likely to show signs of guilt and shame as toddlers.

Grazyna Kochanska, at the University of Iowa, has been studying the relationships among a child's temperament, the mutual responsiveness of mother and child, and the development of conscience in young children. She found that the most effective way to encourage shy, fearful toddlers to adapt to parental standards was through "gentle discipline," or warm, authoritative parenting. For the bolder children, the optimal condition was one in which the mother and child were mutually responsive and cooperative. The child was sensitive to the mother's praise or disapproval and generally showed a desire to be helpful. In general, when the relationship between child and parent was judged to be a "good fit," the child's respect for parental standards remained consistent into preschool age.

How is Steven as a first-grader? We saw him as a somewhat "dreamy" baby, and as a toddler and preschooler he was easily distracted and quickly lost interest, even when he began a project with enthusiasm. His teacher will need some patience before Steven grasps the idea of reading or gets his hand to guide his pencil in forming the letters carefully. Since his mother is aware of his problems, she will limit his distractions and encourage him to finish a job once he starts it. A lot is still going on in Steven's brain, helping him to improve his coordination and to focus his attention.

Sonja's mother saw how persistent Sonja was when her daughter tried to empty her purse at Emily's first birthday party. Sonja continued to follow through on what she wanted to do and push distractions aside. Her unusual musical ability was soon apparent to her parents, who were musicians themselves. We see her as a self-assured toddler and again as a highly motivated violin pupil. She enjoys her new school and has a circle of good friends.

Matthew was an early walker and was well ahead of his pre-school friends in jumping, hopping, climbing, and using a tricycle. At Emily's sixth birthday party, he is a confident, outgoing boy who enjoys sports. He readily takes on the role of leader and doesn't hesitate to speak up to Tommy.

We are more concerned about Tommy than about any of the other children. As a toddler he had a strong tendency to act impulsively and without showing any feeling for others. In first grade he still has difficulty controlling his anger and is often aggressive and destructive. These are all characteristics that long-term studies have linked to later problem behavior. However, it is not a child's temperament alone, but often the mismatch between the child's temperament and the conditions of his environment that increases the probability of later difficulties. The ability of parents to respond sensitively and appropriately to their child's difficult temperament can reduce the risk that their child will be violent and destructive later.

Unfortunately, Tommy's mother is overwhelmed and has very little influence over him. If she can learn to help him cope with his anger and to cooperate with others, he will have a better chance to develop his personal strengths and abilities.

Looking at temperament in adults, C. Robert Cloninger identified four main behavioral traits that can also be observed in children: novelty seeking, harm avoidance, persistence, and reward dependence. A child who is a novelty seeker is eager to explore new places and objects and likes meeting new people. A child who is very shy with new people or afraid to try out a new toy might be said to be a harm avoider. Persistence refers to the ability to stay with a task and not give up easily. Reward dependence here means a warm, close relationship with caregivers and a need for an "echo" from them. Both positive and negative groups of traits arise, depending on the particular combination of these ingredients or on their relative strength. Harm avoidance, for example, can make a child fearful, pessimistic, or extremely shy; or it can help to make him avoid senseless risks. Novelty seeking can result in creative solutions to technical or artistic challenges, or it can lead to thrill-seeking behaviors that are ultimately self-destructive. Research suggests that neurotransmitters play a role in these behavioral characteristics.

Temperament and neurotransmitters

Several neurotransmitters that may play a role in temperament have been tentatively linked to specific types of behavior. Norepinephrine increases alertness and focused attention. Serotonin is involved in moods. High serotonin metabolism has been associated with high impulse control, reduced aggression, and high harm avoidance. Low serotonin activity may increase a person's vulnerability to frustration and contribute to a general negative mood. Very low serotonin activity is reflected by poor impulse control, increased aggression, and risk behavior.

Studies of the neurotransmitter dopamine provide an example of how these messenger substances are related to behavior. Dopamine plays a role in the brain's reward systems. High dopamine activity in special areas of the brain increases the enjoyment experienced during a particular activity, thereby increasing the motivation to repeat the experience. Knowledge of what this neurotransmitter does comes from different types of investigations. In a genetic study, adults who showed a high degree of novelty seeking were found to have a high occurrence of the gene that is responsible for a special type of dopamine receptor. It is interesting that this correlation is found in persons of such different ethnic groups as Ashkenazi and Sephardic Jews, non-Jewish Caucasians, Hispanics, Asians, and African-Americans.

It is important to emphasize that the same neurotransmitter can have different effects, depending on the type of receptor, or docking site, that takes it up. The D-4 receptor is associated with novelty seeking. Therefore, individuals with a large number of these receptors in a particular area of the brain may have a greater tendency to seek novelty. Behavior thus has a genetic basic. However, it is the result of the interplay of many factors. If novelty seeking is tempered by the trait of strong harm avoidance, a person may avoid risk behavior yet still engage in innovative intellectual activity. Whether an activity is experienced as pleasant or disagreeable will influence whether that activity is repeated.

Temperament and the frontal cortex

Research on the specialized role of the two brain hemispheres in processing human emotions provides evidence that certain facets of temperament have a brain basis. Richard J. Davidson, at the University of Wisconsin, is one of many neuroscientists today who are exploring the brain circuitry that underlies emotions.

In EEG studies, Davidson and his colleagues found that the left side of the prefrontal cortex is relatively more active than the right side in processing positive, outward-reaching emotions related to setting and attaining goals. The corresponding right side is more active in withdrawal and negative emotions. Depending on which side is more active at the moment, a person will report more positive or more negative feelings. These differences are also reflected in temperament. People whose left prefrontal cortex is more active generally tend to be more optimistic and extroverted. Those whose right prefrontal cortex is more active may be more pessimistic and withdrawn.

EEG techniques have been used to study the relationship between frontal brain activation and a child's ability to regulate his emotions. Nathan Fox and his colleagues first observed four-year-old children playing together in the group. The groups included children who had been described as very outgoing, or extroverted, and very reticent, or introverted. The play sessions were videotaped, and specialists who were not informed of the purpose of the study made notes on the children's behavior.

Two weeks later, the investigators took EEG measurements over the front part of the children's brains. The results showed a

correlation among a child's reported temperament, observed social behavior, and brain activity. The researchers found four basic
constellations:

1. Extroverts, or highly sociable children, who tended to externalize, or act out their problems and resort to aggressive behavior:
right frontal activation was stronger than left.
2. Extroverts, highly sociable children, who did not externalize
their problems and who were less aggressive: left frontal activation
was stronger than right.
3. Introverts, or shy children, who had a strong tendency to internalize problems: right frontal activation was stronger than left.
4. Introverts, shy children, who showed less of a tendency to
internalize problems: left frontal activation was stronger than right.

The investigators of this study suggest that greater activation
of the left hemisphere is a sign that the brain is taking advantage
of language and analytic skills that require the left hemisphere in
order to handle a particular situation. Greater right frontal activation may mean that the left hemisphere, with its access to these
coping mechanisms, is less able to do its job. It is, therefore, more
difficult, to regulate emotion. The results of the study show that
the children with stronger right activation have less control over
their emotions than those with stronger left activation.

These findings do not mean that a child's behavior is programmed to remain constant throughout life. However, some children may require more than the usual amount of guidance in
learning to cope with their emotions. If you are aware of your
child's temperament, you can help him learn to cope better with
his negative emotions and help him look on the bright side.

Temperament and health
Temperament affects a child's health, and children with highly
sensitive nervous systems are more likely to have health problems than their easy-going peers. This is because the high reacters'
nervous systems have a tendency to switch rapidly into "alarm
mode" and remain there. If alarm periods are short, stress responses soon diminish, and the body returns to its usual state.

But if a child is constantly in a state of alarm, his stress reactions remain high. Over a long period of time, a high sensitivity to stress can increase the risk for disorders such as high blood pressure, circulation problems, or allergies.

The picture of a child frightened by a dog illustrates the interactions between the nervous system and organs of the body during an alarm situation. The child is playing happily in a sandbox. Suddenly he sees a large, strange dog approaching him. Information about the new situation goes from the child's eyes and ears to his thalamus and on to his amygdala. The amygdala is part of the limbic, or emotional, system and plays a major role in temperament. It modulates the strength of incoming signals. If the signals come through strong, the child is likely to be a high reacter. If the signals come through with diminished strength, the child will probably react less intensely.

The amygdala sends its signals to the basal ganglia, causing the child's body to "freeze" and his hand to drop his shovel. The amygdala also sends signals to his brainstem. He becomes more alert. His eyes open wide, his lower jaw drops, and he lets out a high-pitched cry. His heart pounds, and for a moment he holds his breath. These responses are the work of his brainstem and spinal cord, which contain neurons belonging to the autonomic nervous system.

The autonomic nervous system gets its name from the fact that it usually works without any specific, conscious orders. Its goal is to keep the functions of the body in balance, not always an easy task in times of danger. The two partners in the autonomic nervous system are the sympathetic and the parasympathetic nervous systems. The sympathetic system serves an alarm function, the parasympathetic system has a calming effect. While the sympathetic system increases the heart rate, the parasympathetic system reduces it. The sympathetic system steps up intestinal movements, the parasympathetic slows them down.

When the child sees the dog, his amygdala sends signals to a third recipient, his hypothalamus. This organ secretes a hormone that activiates the pituitary gland, and this in turn, tells the adre-

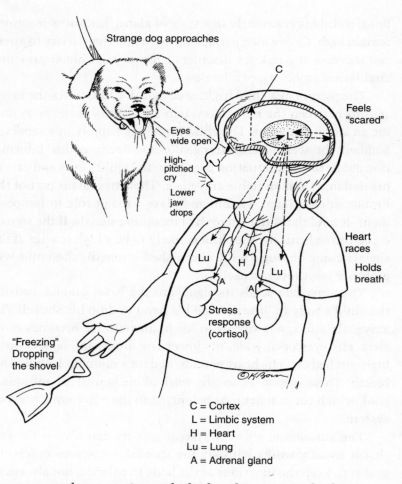

Strange dog approaches

C

Feels "scared"

L

Eyes wide open

High-pitched cry

Lower jaw drops

Heart races

Holds breath

Lu   H   Lu

A   A

Stress response (cortisol)

"Freezing" Dropping the shovel

©K Born

C = Cortex
L = Limbic system
H = Heart
Lu = Lung
A = Adrenal gland

An unexpected event activates the limbic alarm system, leading to emotional body responses. The cortex becomes aware of these responses and translates them into a feeling of being scared.

nal cortex situated on top of the kidney to excrete cortisol, often called a "stress hormone." The reactions of the child's basal ganglia (freezing, startle movements), brainstem (heart rate, breathing), and hypothalamus (cortisol), can be summed up as emotional responses.

Now the cerebral cortex comes into the picture and "interprets" both the event itself and the child's emotional responses to it. When he first saw the dog, his thalamus also sent a message directly to his cerebral cortex, which compares the dog to ones that the child knows and the present situation to previous experiences he's had with dogs. His cortex also receives the input from his emotional responses. He feels his heart pounding, senses the tenseness of his muscles and hears his own frightened voice. His cortex translates his awareness of these emotional responses into "feelings." He feels "scared." In contrast to his immediate emotional responses, his feelings cannot be measured or observed by others. They are private. It is still a great mystery how the brain translates emotional responses into personal feelings.

The cerebral cortex not only receives and interprets information; it also exerts an influence over the subcortical structures, affecting both the child's internal organs and ultimately his behavior on similar future occasions. If the child was badly frightened by the dog, he may be more alert and more upset in future encounters with strange dogs. However, if dog owners keep a firm hand on their animals and encourage the child to approach and pat the animal's fur, with time he may learn to overcome his fears. Because the cortex has the ability to modulate the immediate reactions of the child's amygdala, even a child with a very sensitive amygdala can learn through positive experiences to reduce his anxiety.

The picture of the child frightened by a dog shows the general principle of how the nervous system interacts with other organs of the body during stressful situations. The intensity with which a child's nervous system reacts is individual. In connection with his studies of infants undergoing the heel-prick for metabolic screening, Michael Lewis described three important factors that help explain how a child's particular temperament influences his body's automatic responses to painful events. The first is the threshold, or how intensive a stimulation has to be in order to provoke a response. The second is how fast a reaction can be dampened, and the third is how easily the response can be reactivated. According to this description, a high reactive child, with a very sensitive amygdala, has a low threshold and reacts strongly to a low level of stimulation. The child is easily excited, slow to calm down, and immediately ready to become excited again. The child has high cortisol levels. Evidence suggests that high reactive children have a combination of a more active sympathetic system and a less active parasympathetic system.

In contrast to the high reactive child, the low reactive child shows a high threshold to stimulation, a rapid dampening of the response, and a slow reactivation. This child does not become excited so easily and when excited, he calms down soon. He is less likely to become excited again soon. The child's cortisol levels remain low.

High reacting children like Anna in our story are twice as likely to have hay fever, asthma, and skin allergies as low reacting children. This is because their amygdala sends strong signals to the hypothalamus, causing the hypothalamus to stimulate the production of the hormone cortisol. High levels of cortisol over an extended period of time can depress the activity of the child's immune system, meaning that he has less resistance to infections and a greater risk of allergies.

In addition to causing an increase in the production of cortisol, strong signals from a child's amygdala to his autonomic nervous system stimulate the nerves that contact his inner organs, leading to occasional bouts of indigestion or diarrhea. A constriction of the blood vessels in his body may cause him to complain of headaches. He may frequently feel tense or have difficulty getting to sleep.

Knowledge of a child's temperament has practical consequences because parents can do a lot to help him reduce, prevent, or cope with stress. This can be done by resisting the temptation to avoid all stressful situations, and by interrupting the cycle of high reactivity and greater anxiety. If the child knows in advance what to expect when he gets an inoculation, that it will only hurt for a few seconds, and that it will hurt less when he relaxes and breathes deeply, he will not let the experience upset him as much as if he were taken by surprise. Frustration, annoyance, and anger can also contribute to stress reactions. If your child bursts out over a trivial matter, you can remind him that now he is letting his anger run away with him. Help him take the little "bumps" in life more calmly.

It is important to realize that the impact of an event on a child's response is not merely due to the nature of the event itself but largely to how the child interprets it. Moving to a new house in another town is a traumatic experience for one child, for another it is an adventure. Shy, aggressive, and high reactive children are more likely to perceive events as stressful situations than the less inhibited, less aggressive, and low reacting children. Parents can help a child cope with shyness and find alternatives to aggressiveness. They can also help a child learn to perceive a situation as a challenge rather than as stress. Point out the positive aspects of a new experience, such as moving to a new town: he will meet new friends or have a brand new room of his own.

Fitting into a new group

Fitting into a new peer group is a challenge that children face when entering a new day care center, preschool, or school. Differences in the way children react to this experience are partially related to differences in temperament. Megan Gunnar and her colleagues investigated the relationship between a child's temperament and how easily he adapted to his new group. The investigators observed three- to five-year-old children during the formation of new groups at their preschool facility. They noted the children's behavior and asked parents for descriptions of their child's temperament. To see how the child's nervous system responded to the experience, they measured the cortisol in the child's saliva. An increase in cortisol level is an indication that the child is experiencing some form of stress. The samples were taken during the period when the children were getting to know each other and forming groups and again after the groups had had time to become established.

The investigators identified three groups of children. The first had a high cortisol response on the first day, but once the group had become familiar, their cortisol returned to the usual level and stayed there. These children tended to be outgoing, competent, and popular with their peers. They were probably excited about their new school experience at first, but they soon felt at home and became more relaxed.

The second group showed no increase in cortisol on the first day but elevated responses after the class had become familiar. Maybe they were less excited by the thought of new experiences in the first place, and once they got to their new preschool, they became more anxious. The third group began with high cortisol responses that did not decrease with experience in the group. Perhaps these children were afraid, and it took them a long time to calm down. The children in these two groups tended to be more negative in mood and withdrawn from the group. They also scored lower in measures of attentional and inhibitory control. These children might require extra attention and encouragement in order to find their places in the new group.

Research has shown that a child's individual cortisol secretion pattern tends to remain stable through adolescence. However, a child's behavior can change. Helping a child with a highly reactive nervous system learn to relax not only improves his health but also makes his life more enjoyable. Aim for a good balance between physical activity and rest. Daily physical exercise is important, particularly in the fresh air; this can be playing or helping with work in the yard. If a child is especially tense or suffers from asthma, you should discuss with his pediatrician the possibility of relaxation exercises and learning special breathing techniques. Children's music, dance, and drawing groups offer the opportunity to release tension and enjoy participating with others. Reduce stress at home by spending more time at the table, giving your child a chance to talk about his experiences during the day.

Your encouragement and example are very important in helping your child cope with new situations. You can help your child talk about his fears. A conversation with Anna on her first day of kindergarten might go something like this:

Anna: *I don't want to go to kindergarten.*

Mother: *Why don't you want to go?*

Anna repeats what she said and begins to cry. The conversation doesn't go anywhere.

Mother: *Do you remember how you felt on the first day at preschool?*

Anna: *I cried.*

Mother: *Yes, you cried at first, but then you found all those nice toys, and Mrs. Johnson let you help her in the kitchen. How did you like it then?*

Anna: *Then it was fun.*

Mother: *So how do you suppose it will be now at the new kindergarten?*

Anna begins to smile. She is beginning to see that events she is afraid of may turn out to be positive after all.

A very shy or anxious child may need to get used to just one new group at a time. Wait until he has become accustomed to his new kindergarten class before you start his judo lessons. If a child learns through experience that it is fun to meet new people or that trying new solutions leads to success in solving problems, he will find the courage to face these situations as challenges he can overcome and not as "stress."

Gifts in the Cradle: Multiple Intelligences

Before she was quite three years old, Sonja got very excited when her father took his cello out of the case and started practicing for his concert with the university orchestra. She listened intently and begged to be allowed to draw the bow over the strings herself. Matthew was using whole sentences before some of his friends had said their first words, and by the time he was five he had an extensive vocabulary and an impressive command of language.

Together with a personal set of temperamental traits, each child is born with an individual palette of abilities that also make their contribution to the child's personality. Some of these talents can be seen very early, but some only become apparent around school age, and some don't really blossom until adolescence or early adulthood.

In the early 1980s Howard Gardner introduced the concept of "multiple intelligences" to counterbalance the current overemphasis on the verbal and mathematical abilities tested on standard intelligence tests. His definition of intelligence included the consideration of the value a society places on a particular skill.

Gardner's list of candidates for the status of an "intelligence" came to include the following abilities: linguistic (using language), logical or mathematical, spatial (sculptors, architects), musical, bodily or kinesthetic (sports, handwork), interpersonal (how people interact with each other), intrapersonal (one's own feelings and motivations), and naturalist (understanding of the natural environment). The theory of multiple intelligences is based on the fact that these special skills can appear as separate "packages." A person skilled in music need not, for example, be particularly skilled in drawing. A great poet may not be skilled in managing his finances or getting along with other people. We have examples, too, of "idiot savants," who have above-average skills in a particular field such as music or arithmetic in spite of general mental handicaps.

Gardner proposes that the basis for the special skills, or "intelligences," is a combination of core processes that are present at birth or emerge early in life, basic settings of the nervous system that make one person more sensitive to sound, say, or more dexterous than another. The further development of the talent involves input from other sensory systems, personal traits such as persistence and involvement, and the appreciation and support of the child's society.

Gardner's theory of multiple intelligences has important consequences. One is that individual strengths and weaknesses should be evaluated in a realistic setting, not just in isolated test situations. For example, he has suggested using a child's "portfolio," a comprehensive collection of the child's work—both in school and during free time—as a means of measuring the child's progress in school. Another consequence is that individual strengths can be used to motivate and support children's learning in other domains. For example, a child who has no interest in reading but who is very good at sports might be encouraged to enjoy reading by introducing him to stories about his athletic heroes.

Many scientists apply a different approach to the question of intelligence. They break up the complex term *intelligence* into single behaviors that can be linked to basic neural processes. Some examples of behaviors associated with a specific neural process are novelty detection, neural processing speed, habituation, memory capacity, and inference. While studies exist comparing results on tests of these basic behaviors to later achievement in school, it would be misleading to assume that they measure the whole quality we think of as *intelligence*.

David Wechsler, famous for his intelligence scales, carefully defined intelligence as a general quality that included traits such as "persistence, zest, impulse control, and goal awareness"— qualities that are not directly assessed in intelligence tests and yet have a great impact on a child's achievements. These are all qualities that parents can influence, both by example and by direct encouragement.

Musical genius and the brain: the example of Mozart

Our fascination with the rare phenomenon of child prodigies leads us to speculate about the underlying reasons for the extraordinary giftedness of Mozart. The notes of one of Mozart's contemporaries are revealing.

Dr. Samuel Tissot (1728–1797), a well-known doctor in Lausanne and a relative of the famous watchmakers, was so impressed by hearing the young Mozart—officially 9 years old, but most probably 10 at the time—that he wrote a detailed summary of his observations in an article of the journal *Aristide ou Le Citoyen* dated October 11, 1766. He suggested that young Wolfgang's great musical ability was partly due to the special nature of his brain. The way his brain processed sound programmed him to be more attentive to sounds in the first place. It made him more sensitive to pitch and harmony. But there was more to it than just the particular sensitivity of Mozart's system for processing sounds. He was drawn "as if by a secret force to his clavier and lured tones from it that were the living expression of the ideas that occupied him at the moment." Music had the power to move him in a very deep and unique way. Mozart's affinity for music led him to spend a great deal of time practicing and this, in turn, had its effect on the fine-tuning of his system.

Dr. Tissot suggested that Mozart's musical genius might never have developed if he hadn't been born into a family where music was a part of daily life. Although the boy's brain had especially sensitive and well-coordinated systems for hearing and fine motor control, practice and discipline were essential. His father's confidence in his ability and his continuous support and encouragement were very much part of Mozart's success. The authors of a recent report on the relationship between "talent" and "deliberate practice" also emphasize the role of parents in encouraging a child's confidence in his abilities and persistence in achieving higher goals.

Some of Dr. Tissot's observations may be relevant for parents today. Tissot reports that Mozart's father didn't have to urge his son to occupy himself with music. On the contrary, he did his best to keep his energies in balance and channel them into efforts that would take him further. Both Mozart and his sister, Nannerl, looked to their father as a model and were very receptive to his praise or disapproval. The court trumpeter, Andreas Schachtner, who knew Mozart as a child, described him as a very lively, enthusiastic child and remarked that he might have turned out badly if he hadn't had such a favorable upbringing.

The phenomenon of Mozart's genius is still a mystery and likely to remain so even when we know more about the brain structures involved in musical talent. Many wunderkinder who start out like Mozart sink rapidly into oblivion. Other composers, such as Camille Saint-Saëns or Felix Mendelssohn, both of whom might perhaps have even overshadowed Mozart in early childhood, did not go on to produce music that competes with Mozart's.

In the attempt to find differences in the brain that would explain the extraordinary achievements of highly gifted individuals, scientists have investigated the brains of Lenin and Einstein. However, their results were not conclusive. If there are differences in the brain that explain exceptional giftedness, we are not likely to find them by anatomical studies of a brain that is no longer alive. It is possible that future research using new imaging techniques will help explain the phenomenon of genius by investigating the biochemical and electrophysiological processes taking place in the living brain.

Nature–Nurture

The exciting news reverberated simultaneously through the scientific world and the popular press. The media even went so far as to speak of an "intelligence gene." A team of scientists at Princeton, Massachusetts Institute of Technology, and Washington University announced that they had altered a single gene to produce a strain of mice that were more successful than control animals in finding their way through a maze, an animal test of memory and learning. The results of these experiments, together with the news of the mapping of the human genome, have rekindled our interest in the age-old discussion of the intimate relationship between nature and nurture.

Research during the last few years has uncovered some of the secrets of how nature and nurture interact. One aspect of this collaboration is the influence of activity on the shaping of brain structures, a process that has been shown to take place at the level of the cell. Each cell contains a copy of a person's individual genetic code, a sequence of nucleic acids that provide the instructions for building the proteins used to make the cell and cell products. Genes are, therefore, a code for proteins, not for behavior itself. Before the instructions contained in the genes go into action, they have to be unlocked and "expressed." Here activity plays a role.

Activity stimulates the release of neurotransmitters. When a neurotransmitter molecule docks at its special receptor on the surface of the cell, the signal carried by the neurotransmitter is whisked from the cell membrane into the nucleus and to the genes. In an intricate process, the gene is activated, its message is expressed, and proteins are synthesized. The proteins are then used as building blocks for dendrites, synapses, or enyzmes that regulate metabolism. This is the place where nurture interacts directly with nature.

The fact that the human brain develops over such a long period after birth allows the formation of an extremely complex network, which is adapted to the activity and needs of the individual and is the basis for the immense plasticity of the brain. A slight mutation of a gene way back in evolution may have been critical in the timing of the developmental processes and led to this extended growth period.

However, it takes far more than a single gene to account for the neural basis of complex traits, such as verbal ability or persistence. These are the result of multiple genes acting together. Each single gene in the group is necessary for the behavior, but it is not sufficient to explain the behavior by itself. Science is just beginning to illuminate some of the relationships between these genes and specific traits, such as language deficits. One of the many genes involved in temperament has received particular attention lately. Individuals with longer strands of a special dopamine receptor gene are more likely to show high novelty- or thrill-seeking behavior than those with short strands of the gene, who tend to be more deliberate and orderly.

While we can't ignore the contribution of genes, we cannot separate them from the effects of the environment. When asked about the relative importance of genetics and environment, or nature and nurture, the neuroscientist Donald O. Hebb countered with the question: "What contributes more to the size of a rectangle: the length or the width?"

The interaction between genes and environment has been extensively explored in studies of twins and adopted children. Since identical twins share 100 percent of their genome and, in a way, are natural "clones," following their development, particularly when they have been separated by adoption, provides important insights into the effects of the environment. Fraternal twins and other siblings share on average only half their genes. The overall results of these investigations reveal that the contributions of genetics and environment to behavioral differences among individuals are about fifty-fifty. However, the importance of the genetic contribution is not equal for all behaviors and may vary at different ages.

On the basis of twin studies comparing identical and fraternal twins, Robert Plomin and his colleagues estimated that genetics accounts for about 60 percent of the variance in verbal ability and about 50 percent of the variance in spatial ability. Verbal and spatial abilities are two of the most heritable cognitive abilities. Verbal reasoning, general intelligence, and extroversion have a higher degree of heritability than either memory or processing speed.

Research suggests that while some components of temperament are influenced by a child's genes, the genetic contribution may be modest when all other factors are considered. A child who has a temperamental bias for shyness may, with experience, come to enjoy social encounters. Shyness, boldness, a sense of humor, curiosity, persistance, and the multitude of other characteristics we associate with temperament are not the result of single systems acting alone. Instead, these systems constantly interact with each other, influenced by both maturation and the child's experience. In general, the genetic influence is higher with respect to extreme forms of behavior.

The staff of the Institute for Behavioral Genetics at the University of Colorado kept track of pairs of same-sex twins from the age of 14 months through late childhood and found evidence of genetic influence on whether a child was an extrovert or an introvert. Extreme shyness was more likely to be shared by identical twins than fraternal twins. For adopted siblings, who are genetically unrelated but live in the same home, the similarities in temperament were no more than chance.

Three large studies compared twins with respect to the same three temperament dimensions at the end of the second year: positive, outgoing demeanor; energy shown in activity; and attention and persistence in goal-directed tasks. The genetically identical twins were the most similar, the adopted siblings (genetically unrelated) least similar. The fraternal twins and biological siblings were in between.

The current research in genetics is actually telling us more about the role of the environment, or as Robert Plomin so aptly puts it, the "nature of nurture." If genes account for about 50 percent of the differences in heritability of temperament, what mysterious factor accounts for the rest? Why are two children in the same family often as different from each other as a pair of children picked at random from the population?

An explanation that applies to all children, not only twins, is that living in the same family does not mean living in the "same environment." Children growing up in the same household experience both shared and nonshared features of their surroundings, and these have an effect on the course of their development. The shared aspects include the culture of the surrounding community, parental education level, and family traditions. However, current research is now focusing also on the nonshared features.

Siblings, even identical twins, growing up in the same household, experience a radically different environment from the very beginning. Even in the uterus, identical twins do not have a completely identical environment. Later, each child has unique personal relationships within the family, both with parents and other siblings. Each child has his own personal way of interpreting events, of choosing his friends and his reading material. Of the 50 percent total environmental influence, the nonshared environment may actually account for as much as 47 percent. This means that the forces that make siblings different can be stronger than those that cause their resemblance.

To Think About

How can knowledge of a child's personality affect parenting?

A concept that is best described by the word *matching* proposes that your child's personality has the best chance to unfold optimally when your expectations, needs, and possibilities are in harmony with your child's temperament, motivation, and abilities. Achieving this harmony is made more complicated by the fact that parents are individuals too. It is important to take all these factors into consideration.

You can make adjustments in your parenting style according to the temperament of your child. A warm, authoritative parenting style is particularly beneficial for children who are shy or fearful. Children with a particularly sensitive nervous system may need a more tranquil and predictable environment than children who are better able to cope with excess stimulation.

Allow time for your child's special strengths to unfold, and be ready to adjust your expectations accordingly. Try not to be disappointed if he shows no interest in the xylophone you bought him or if he doesn't find reading as easy as you did when you were his age.

Parents may find themselves orchestrating family life with children who have very different temperaments and abilities. Let's suppose a couple has a three-year-old son, Adam, who is rather shy and serious. He reaches all the motor milestones on time, but he is slow and he has to concentrate in order to make his fingers do what he wants. Now his baby brother, Ethan, arrives. From the very beginning Ethan charms the world around him with his smiles. He's an early walker, an early talker, and a lively little fellow. Adam may well feel left out and may become discouraged when he sees that his little brother learns so effortlessly. Adam's parents can help him discover and develop his own special strengths and interests. They can help him find playmates with whom he feels comfortable.

What can I do about a child who has frequent negative moods?

For some people the glass is half full, for others the same glass is half empty. Although basic moods have a biological basis and are part of temperament, they can be influenced by experience. If your child frequently uses words like, "I'm never any good at this," help him to see new ways to go about solving a problem. Point out activities where he is successful. Encourage his persistence and help him overcome frustrations. Try to project a positive attitude and help booster his self-confidence by making him an active participant in the family team.

What should I do if my child is an introvert?

Somehow we get the feeling today that everyone should be outgoing, exuberant, a leader. This would be a boring world, not to say an exhausting one. Your child doesn't have to be the life of the party. He can have a few good friends. Introverts are often more cautious and reflective than their more outgoing peers. They can have a lively imagination and be very creative.

However, introverts sometimes have special problems. If your child is painfully shy, you should help him make positive experiences with other people because this will make it easier for him later to be part of the group. Some shy children tend to blame themselves when things go wrong and to take things too much to heart. Learning to talk about these feelings may help a child counteract a tendency that could otherwise lead to later depressive symptoms. An extremely small number of shy and withdrawn children bottle up their anger and resentment inside, accumulating hostility that can erupt in violent and brutal acts. It is important to identify these children early and help them learn to assert themselves and solve conflicts in positive ways.

Ten Guideposts
for Parents

In today's rapidly changing world, it's no wonder if you sometimes feel as if you had landed unexpectedly in the middle of a dense forest called parenthood. Your map is confusing, you don't have a guide, you don't even know where you're going. We have extracted this list of guideposts that we think represent important behaviors from material in the previous chapters of this book. These are behaviors that normal brain development puts within the reach of the five- to six-year-old child all over the world, whether she grows up in a residential suburb, an inner city, a farm, or in a fishing village.

Because we feel the guideposts represent behaviors that are of crucial importance, not only in childhood, but all through life, we were not at all surprised to discover some of them among the factors that George E. Vaillant describes in his recent book *Aging Well: Surprising guideposts to a happier life* from the landmark Harvard Study of Adult Development. The capabilities of dealing with stress, of enjoying lifelong learning, and of maintaining close

> At six years of age a child can
>
> 1. form personal relationships
>
> 2. participate in dialogue
>
> 3. show empathy
>
> 4. exercise a sense of fairness
>
> 5. set goals and try to reach them
>
> 6. enjoy creative activity
>
> 7. learn from experience
>
> 8. overcome frustrations
>
> 9. take responsibility
>
> 10. take wider perspectives

personal relationships can all grow from the foundations established in childhood.

The 10 Guideposts

Think of each guidepost as a marker representing one central feature of a child's behavior. This list does not imply a particular hierarchy, and we could well imagine listing the guideposts in a different order. Maybe you will add new ones to the list. Of course, these guideposts do not mean a child must reach the same level of the particular behavior that would be expected of a teenager or of an adult. But they should give you an idea of the direction in which your child is heading. Keeping in mind the wide range of individual variation, you will become aware of when your child's "compass" needs adjustment.

The fact that children are capable of the behaviors illustrated by our markers does not mean that they will necessarily develop the skills all by themselves. Guidance from the child's social surroundings is essential. This is similar to the way children learn their native language. They are born with the capacity for language, but experience determines which language they will speak and to a great extent how well they will speak it. To demonstrate some suggestions for using the guideposts, we call in the fictional parents and children you have already met.

Guidepost One: Personal Relationships

Deborah and Allen are looking at guidepost number one, the nature of a child's close personal relationships. Close personal relationships are something we often take for granted. Children are born into the world completely dependent on adults for their care. As they grow, their relationships become less one-sided and more mutual. Mutuality encourages respect and cooperation and reduces the need for disciplinary measures. Just saying, "I love you," is not enough.

A good relationship has to be lived and felt in countless interactions in daily life. Emily is very outgoing in her display of affection. She bounds down the stairs in the morning and wraps her arms around her father's waist. She can hardly wait to tell her parents about what she did in school. But Emily demonstrates her relationship to her family in other ways as well. When Deborah cut her finger Emily rushed to bring her a Band-Aid and offered to help wash the dishes. Emily likes working with her parents in the garden or the kitchen and loves to hear a story in the evening. Emily openly admires her father and often imitates his expressions or repeats his jokes. Emily is very fond of her grandparents and often spends the weekend with them.

Deborah makes it a point to know Emily's friends and their parents. This is important, because Emily is now spending more time outside the home and bringing home new ideas. Deborah takes time to sit and talk with the children while they are eating cookies or sharing a pizza in the kitchen. This gives Deborah a chance to find out what the other children are interested in. Through her friends Emily experiences the give and take of friendship.

Although Andrew is younger than Emily, the thought occurs to Deborah and Allen to ask themselves the same questions about him. Andrew is much less outgoing than Emily, so they have to think for a few minutes. Does he enjoy talking and being together? Does he participate in family traditions, and does he cooperate in family tasks? Does he have friends? Andrew does like to look at books with his mother or father. He helps willingly when asked, but he doesn't help spontaneously like Emily. He isn't immediately comfortable in a group of children, but he has one good friend that he likes to play with. Until recently he wasn't very good about sharing his toys with his friend, but now he is learning that sharing belongs to friendship.

Deborah and Allen are happy with their children's choice of friends. Peer groups are becoming more important. But they do not alone determine a child's behavior. Parents should observe the kind of friends their child seeks. They may need to point out, for example, that he or she can refuse to go along with the group in any action that would cause harm to another person.

Guidepost Two: Participation in Dialogue

"Of course he talks," Steven's father says. But the guidepost poses a more specific question: "Can he participate in dialogue?" Steven sometimes doesn't pay any attention to whether anyone is listening to him or not. In fact, he sometimes doesn't even hear their questions—if they manage to insert them into his flood of words.

Steven's parents decide to do something to help him become an active participant in conversations. They take advantage of situations that occur in daily life: mealtimes, riding in the car, cleaning up the kitchen together, or story time. They give Steven their full attention when he is talking to them. If he starts to reel off his narrative at full speed, they might say, "Now wait a minute, I think I missed something. Where did you say the cat was hiding?"

Communication is a two-way street, and Steven also has to learn to listen to others. One way to help him learn is to set a good example. Steven's parents make sure that everyone in the family has a chance to speak at the dinner table. They keep long discussions about their work for later when they are alone. They encourage him to use persuasion with his little brother rather than physical force.

When Tommy's mother stops at the language marker, she has to admit that Tommy doesn't really listen or care about what other people are saying. He is more likely to push her aside than to ask her nicely to let him pass. He expresses his arguments on the playground with his fists. It would be time to have a talk with his teacher.

Tommy's mother realizes that her son needs more time to practice speaking and listening. She will get up a few minutes earlier so she has time to sit and eat breakfast with him. Breakfast gives a child a good start for the day. It is a time to collect one's thoughts. But also, good nutrition is important, especially for children.

She can try taking him on excursions to places that interest him—the zoo, for example. Here she can give him her full attention and ask questions that he can answer because he seems to know more than she does about the animals there.

Guidepost Three: Showing Empathy

Empathy is the ability to participate in another person's emotions and feelings, to share their worry, pain, or joy. Empathy is the foundation for caring social behavior, and a basis for fairness and a moral sense. During the course of their second year, as children become aware of themselves as individuals, they expand their attention from their own concerns to those of others around them. They try to comfort and offer their help. By the age of six, they can discuss their feelings and those of others and can feel empathy for people who are not present. These can be characters in a story or in a film. It is important that they learn empathy not only for their own particular group.

Matthew's parents made a note of some of the situations in which their son showed empathy. When his mother had a headache one night, Matthew offered to make dinner so she could lie down for a while. He comforted his little sister when she stubbed her toe. He brought home a bird that had fallen out of its nest and tried to make it comfortable by fitting out a cardboard box with some leaves and twigs.

But what can parents do if they have the feeling their child does not show empathy? Children who appear to be uncaring or who have difficulty sharing can be encouraged to do so by appealing to their pride and desire to be grown up. However, reason by itself is not enough. And rewards are ineffective. A child must have the opportunity to see adults showing empathy with others in daily life. Parents can reinforce their example by talking about the feelings of others. Stories are a good way to point out situations where a character showed empathy.

A child should see how pleased people are when she does something for them. Be sure to thank her for the flowers she brought when you were sick. She should see how pleased people are when she does something for them.

Shy children may show their empathy less than outgoing children do. This does not necessarily mean that they are less concerned. They may be too occupied with their own feelings of distress or they may be afraid to reach out to other people. Parents can help them by talking to them about their feelings. Anna was afraid to visit Emily in the hospital when she had her tonsils out, but when Anna's mother calmed her fears about the big hospital and told her how happy Emily would be to see her, Anna went to visit her friend. And she saw for herself how happy it made Emily.

Guidepost Four: Showing a Sense of Fairness

Most of the children in our story showed that they had developed a concept of fairness that was more than just abiding by the rules. Like the children who heard the story about the two boys in the doctor's waiting room, they felt that the boy who had been waiting longer had to let the other boy get treated first because he was bleeding and in pain. The children knew that the concept of fairness means being able to look at a problem from different angles and then acting accordingly. Being fair includes both thinking and feeling.

We saw an example of how children learn to interpret more than one dimension of a dilemma in the story of the witch and the grandmother, which we introduced in the section on overcoming prejudice. The four-year-olds always thought the "grandmother" was good, no matter how she handled the cat that ventured into her kitchen. The six-year-olds were showing fairness by judging the witch and the grandmother on the basis of their actions and not on the basis of appearance alone.

William Damon investigated the development of children's sense of fairness by telling them a story and asking a series of questions. He told them that a group of children spent a whole day drawing pictures. Some made more, some made fewer. Some drawings were good, some drawings were not so good. Some of the children worked hard, while others fooled around. Some of the children were poor, and some were rich. Some were boys, and some were girls. The pictures were put up for sale at the school bazaar.

Then Damon asked the children who heard the story how the proceeds from the sale should be fairly distributed. The children under four simply stated their own desires. The four-year-olds also stated basically what they wanted themselves, but they began to offer some arbitrary reasons. The five- to seven-year-olds suggested that all participants had a claim to the rewards and the best way to resolve the conflict would be to give everyone an equal share. From the age of eight, children began to consider individual circumstances. Perhaps children should get more than the general share if they contributed more or if they were handicapped by poverty or physical disability.

Damon went on to see whether the eight-year-old children would actually act the way they had indicated after hearing the story. He gave them the task of distributing candy bars among themselves as payment for their work. Fifty percent of the children behaved the way they said they would. About 10 percent of the children actually exhibited more advanced reasoning than they had shown in the theoretical situation. The remaining 40 percent of the children did not put the standards they had set themselves after hearing the story into practice. Other researchers have obtained similar results using tasks like the one Damon used.

The parents in our own story can bring many examples of how their children show a consideration of the needs and rights of others. The ones who have seen Tommy on the playground may have their questions about him. But let's hope that Tommy's mother will remember this guidepost next time she watches him at play.

Guidepost Five: Setting Goals and Trying to Reach Them

Our little group of parents looks puzzled at guidepost number five. What kind of goals do six-year-olds set? After thinking for a while, Emily's mother suggests that this could be trying to read all the books by a child's favorite author. Sonja's father adds that it might mean preparing for a recital. It could also mean finishing a tunnel for an electric train. Emily's father thinks it could be the wish to learn a lot about something, for example, dinosaurs. A child might go to the museum, find out what dinosaurs ate, draw a series of pictures of dinosaurs. Steven's father says saving up for a bicycle is also a goal a child might set for himself.

All these things can be thought of as "goals." In each case, the children set themselves a task and go about finding ways to reach it. In the example of the bicycle, Steven might plan to save his allowance and put it together with some money he got for his birthday. He is planning ahead and ready to postpone instant gratification in order to reach his goal. The other goals involve performing a specific activity.

The parents see the important things their children are learning. The children concentrate on what they are doing because they really enjoy it. And because they have a goal in mind they accept some of the tasks that are not so much fun. Although practicing the violin is not always exciting, Sonja knows that it is important. Emily complains a bit when she has to wash her paintbrushes and clean up the kitchen table, but she knows this is part of the job.

Most of the parents are happy that their children not only enjoy a sense of accomplishment when they finish their projects but that they also have fun planning them and doing the work. The children are learning that their own creative activity is exciting. They would rather do things by themselves than watch television. Of course, they also enjoy their favorite programs once in a while.

Steven's parents have noticed that Steven has a habit of starting projects and then not finishing them. They will help him learn to try new ways to solve his problems and encourage him to finish one thing before he starts another. By helping him stay with a task longer, they will also be helping to reduce his tendency to become easily frustrated.

If a child avoids setting goals it might be because she has become discouraged. Perhaps she sets her goals too high and therefore is constantly disappointed. Parents can help a child set realistic goals that are still a challenge.

Guidepost Six: Enjoying Creative Activity
The sixth marker has to do with curiosity, imagination, and taking pleasure in art and music. These qualities not only help children learn about the world and develop new ideas but also help them to a deeper understanding and enjoyment of life. The parents in our story tried to come up with examples of their children's behavior that might illustrate the meaning of the signpost. Sonja enjoys her music. Emily likes to make animal figures out of clay. Steven likes to sing and dance in kindergarten.

Anna's mother says that Anna has always enjoyed listening to stories and that she often pretends she is one of the characters in them. Anna likes to ask what-if questions like "What would I do if I were the teacher?" She enjoys playing a role and taking another person's point of view. Her imagination allows her to explore her world and make the acquaintance of new people even if she is not really the adventurous type.

Tommy's mother suddenly realizes that Tommy spends almost all his free time in front of the television set. She remembers that the drawings he brought home from kindergarten showed a lot of imagination. Maybe she could reduce his TV hours and encourage him to paint a story in pictures. She also remembers the announcement in her local paper about a children's drawing group at the art museum. It would be worth a try.

Children all over the world take pleasure in painting, in music, and in dance. Active participation in the arts opens new worlds for a child to explore. These stimulate a child's imagination and provide opportunities for her to experience the joy of creativity and to express herself in modes other than speech. Because of the universal human elements expressed through the arts, the French neuroscientist J.-P. Changeux has suggested that they may be the key to promoting understanding among the many different cultures of the world.

Guidepost Seven: Learning from Experience

Daily life is full of opportunities for children to learn that their actions have consequences. If they are capable of learning this, they will be more likely to take responsibility for their actions in the future. A general awareness of consequences will make them less likely as teenagers to endanger their health by smoking or taking drugs.

Steven's mother immediately thinks of a chance she missed just that very afternoon. Steven ordered chocolate ice cream, and when it arrived, he refused to eat it because he now wanted strawberry like the child next to him. Steven's mother gave in to him and gave him her dish of strawberry ice cream. This certainly didn't teach him that his actions have consequences.

In situations that are not dangerous, give your child a chance to make a mistake and feel the consequences for himself. When Matthew went to visit his grandparents over the weekend, he in-

sisted on packing his suitcase by himself. He thought of his toothbrush, his socks, his shirt, and a sweater. He told his mother he
didn't want to take a raincoat. When it rained and he had to stay
inside all afternoon, he was sorry he hadn't taken her advice. The
next time he went away he took his raincoat with him. He had
learned his lesson.

Guidepost Eight: Overcoming Frustration

Since frustrations are a fact of life, a child who has learned to
overcome frustrations rather than simply avoid them is better
equipped for the future. Such children learn persistence and don't
let an occasional failure get them down. Adolescents frequently
list boredom and frustration with life as reasons for taking drugs.

Steven's mother told a story about the bicycle they gave
Steven for his birthday. He dashed out proudly, all ready to head
down the street to join his friends. But when he tried it out, he fell
down. He tried again with no success. Although he wasn't hurt,
he felt discouraged. He put the bicycle in the garage and didn't go
near it. He had given up. Steven needs to learn now in daily life
situations that he can overcome his frustrations or he will avoid
similar situations in the future. This will prevent him from trying
out new things. Steven's father says he will help him practice
when the boy's friends are not around. He may have to explain
that not all boys take to a bike like a duck to water and that he
had to practice a bit, too, when he was Steven's age. When Steven
does manage to ride down the driveway, his success will teach
him that his efforts were worthwhile. This is also an example of
learning from experience.

Coping with emotions is part of learning to deal with frustration. Sonja occasionally "makes a scene" when she doesn't get
her own way. One evening, when she lost the game, she tossed
the Parcheesi board on the floor, kicking it and calling it a stupid
game. Sonja's mother knows there is no point in telling children
they shouldn't be angry, because they often have very understandable reasons. However, she makes it clear that harming other
people or animals or causing wanton destruction is not the way to

resolve one's anger. The next time Sonja is angry, her mother reminds her of the time *she* was very, very angry because someone backed into her car at the parking lot. Although she was very upset about it, she didn't go around kicking all the other cars. This thought makes Sonja smile. Kicking the Parcheesi board begins to look like a silly way to solve a problem. Being able to laugh about a situation is a good way to relax and see it in its proper perspective.

Guidepost Nine: Taking Responsibility

In many parts of the world, children—often younger than six years old—take care of younger siblings and perform tasks like bringing water, washing, running errands. They do not have to be told all the time how wonderful they are. They know they are needed, and this gives them a sense of importance. In an agricultural society with few technical aids, it is easier to find jobs for children of all ages. In our modern society, it may take more imagination to combine the chance to be useful with opportunities for learning and for play. However, we should not underestimate our children. They can make a real contribution to family life.

In order for children to develop a sense of responsibility, it is important for them to have regular chores—and for extended periods. This helps them to learn not only that their efforts are appreciated but also that their chores, though perhaps not interesting in themselves, are part of a larger purpose. As more parents work outside the home, they have less time to devote to the necessary tasks involved in running a household. As children get older, they can take over more responsibility. For example, a three-year-old can set the table. Four- and five-year-olds can make salad. As your child gets older, she can take over planning and preparing a meal for the family. Six-year-olds can sort laundry, carry out garbage, run a vacuum cleaner.

Helping in the home should be motivated by a sense of importance, not by financial rewards. An article on "tweens" (ages 8–14) in the October 18, 1999, issue of *Newsweek* reported that younger and younger children are getting paid for household jobs such as preparing meals, doing gardening work, or working on the computer for their parents. In doing so, they generate income in the billions of dollars. But this trades off crucial learning in family life for a less valuable lesson. In order to learn about money children should receive a regular amount of pocket money, not to be replenished early if the child wastes it.

Guidepost Ten: Taking Wider Perspectives

By the time they are six, most children are capable of taking an interest in more than their immediate surroundings. They are able to think beyond their own concerns to those of a larger group. Their circles are widening to include membership in a particular ethnic or cultural group, and they gradually absorb the standards and values of this society. It is important that they learn that other groups or societies exist and that one is not better than another. Many teachers invite members of other communities or visitors from other nations to spend a few hours in the class, giving the children a chance to meet them and ask questions.

Children are learning that they can help other people without expecting a reward. They often show great concern for people who are disadvantaged or ill. A child may ask about a homeless person she sees on a busy city street. Without overwhelming your child with long explanations, you can encourage her sensitivity for social issues.

Matthew wanted to know why a blind man used a stick, so Matthew's mother put a blindfold over his eyes to give him an idea of what it meant to be blind. She was also lucky to have a blind neighbor, who talked to Matthew about using his stick. The man was anxiously waiting for his new seeing-eye dog and promised to invite Matthew over as soon as the dog got settled in its new home.

Talking to grandparents, great-grandparents, or other members of the older generation gives children a sense of the needs of older people. At the same time, it shows them that these people often have very interesting stories to tell.

The children in our story had all heard of the importance of the natural world around them. They knew that fish needed clean water and that wild animals needed fields and large forests where they could find food and a safe place to live with their families.

From now on, a child gradually spends less time under direct adult supervision and spends more time with people outside the immediate family. Ever widening circles will contribute their influence on the child's growing mind. Developing the basic skills discussed in these guideposts gets your child off to a good start.

Glossary

Amygdala Brain structure involved in emotional responses. Part of limbic system.

Astrocyte Cell that delivers "fuel" to the neurons from the blood, removes waste from the neuron, and modulates the activity of the neuron.

Autonomic nervous system Controls functions of internal organs, e.g., blood pressure, respiration, intestinal motility, urinary bladder control, sweating, body temperature. Its actions are mainly involuntary. Consists of the sympathetic and the parasympathetic nervous systems.

Axon Elongated outgrowth of nerve cell body, sends information to target; also called nerve.

Basal ganglia Subcortical structure involved in motor, cognitive, and emotional functions.

Brainstem   Subcortical structure that connects the cortex, subcortical structures, and spinal cord; coordinates autonomic functions and arousal.

Broca's area   Cortical area involved in language comprehension and in the preparation of muscle movements for speech.

Central nervous system   Brain and spinal cord.

Cerebellum   Modulates force and range of movements, involved in learning motor skills and further in cognitive and emotional functions.

Cerebral cortex   Outer layer of the brain containing the cell bodies of the neurons (gray matter). The cortex is divided into specialized primary, secondary, and associational regions of sensory and motor areas.

Corpus callosum   Bridge linking the two hemispheres of the brain

Dendrite   Branchlike "antenna" of the neuron; receives information.

Electroencephalogram   Technique for recording electrical activity of the brain; referred to as EEG.

Event related potentials   Electrical activity of the cortex in response to stimulation; referred to as ERP.

Gestation   Refers to the time the embryo or fetus spends in the uterus.

Hippocampus   Brain structure involved in memory and learning.

Hypothalamus   Subcortical structure that regulates autonomic, endocrine, and immunological functions.

Limbic system   Involved in all forms of emotional behavior and thinking, in learning and memory formation, and in emotional responses of body and autonomic system. Comprises the amygdala, hypothalamus, hippocampus, and parts of the thalamus and cerebral cortex.

Magnetic resonance imaging Technique based on magnetic fields; used to produce pictures of brain structures without using X rays; referred to as MRI. Functional MRI (fMRI) shows the region in which special activity is taking place.

Melatonin Chemical substance involved in sleep–wake cycles.

Myelin Insulating sheath around the axon that increases the speed of transmission of impulses along the nerve.

Neuron Nerve cell.

Neurotransmitters Chemical messenger substances that connect neurons at the synapse; examples: dopamine, glutamate, gamma-amino-butyric-acid (GABA), serotonin, epinephrine.

Oligodendrocyte Cell that produces myelin.

Peripheral nervous system The nervous system outside the brain and spinal cord.

Pineal gland Brain structure associated with mechanisms that regulate circadian rhythm.

Pituitary gland Brain structure that stimulates other glands to produce hormones involved in body growth, lactation, sexual maturation, stress, and the regulation of general metabolic processes. Secretion of hormones by the pituitary gland is controlled by the hypothalamus.

Positron emission tomography Technique for producing images of functioning brain areas using small amounts of rapidly decaying radioactive substances; referred to as PET.

Receptor Docking site for neurotransmitter on the receiving side of the synapse.

REM sleep   Sleep characterized by rapid eye movements; associated with dreams.

Reticular formation   Central core of brainstem; receives information from most of the sensory systems and regulates attention.

Subcortical structure   Brain structures located underneath the cortex.

Synapse   Connection between nerve cells.

Thalamus   Main relay center between the cortex and the subcortical centers. Entrance to the brain for information from the senses. Has connections with motor, association, and limbic cortical areas.

Vestibular system   System for maintaining balance and equilibrium.

Wernicke's area   Association area in cerebral cortex involved in the comprehension of language.

White matter   Area under the cortex containing the myelinated axons of neurons.

References

For Further Reading

Astington, J.W. 1993. The Child's Discovery of the Mind. Cambridge: Harvard University Press.

Barnet, A.B. and R.J. Barnet. 1998. The Youngest Minds. New York: Simon & Schuster.

Brazelton, T.B. 1992. Touchpoints. Reading, Mass.: Perseus Books.

Bruer, J.T. 1999. The Myth of the First Three Years. New York: Free Press.

Chess, S. and A. Thomas. 1989. Know Your Child. New York: Basic Books.

Damon, W. 1988. The Moral Child. New York: Free Press.

Damon, W. 1996. Greater Expectations. New York: Free Press Paperbacks.

De Loache, J. and A. Gottlieb. 2000. A World of Babies: Imagined Childcare Guides for Seven Societies. Cambridge, UK: Cambridge University Press.

Eisenberg, A., H.E. Murkoff, and S. Hathaway. 1996. What to Expect: The Toddler Years. New York: Workman Publishing.

Eliot, L. 1999. What's Going on in There? New York: Bantam Books.

Gardner, H. 1991. The Unschooled Mind. New York: Basic Books.

Goleman, D. 1995. Emotional Intelligence. New York: Bantam Books.

Gopnik, A., A.N. Meltzoff, and P. Kuhl. 1999. The Scientist in the Crib: Minds, Brains, and How Children Learn. New York: William Morrow.

Greenspan, S.I. 1997. The Growth of the Mind. Reading, Mass.: Addison-Wesley.

Harris, J.R. 1998. The Nurture Assumption. New York: Touchstone, Simon & Schuster.

Kagan, J. 1984. The Nature of the Child. New York: Basic Books.

Kagan, J. 1987. The Emergence of Morality in Young Children. Chicago: University of Chicago Press.

Kagan, J. 1994. Galen's Prophecy. New York: Basic Books.

Karmiloff-Smith, A. 1995. Baby It's You. London: Ebury Press.

Postman, N. 1994. The Disappearance of Childhood. New York: Vintage Books.

Small, M.F. 1998. Our Babies, Ourselves. New York: Anchor Books.

Sternberg, R.J. 1996. Successful Intelligence. New York: Simon & Schuster.

Selected References by Chapter

Chapter 1: Life in the Womb

Eskenazi, B. 1999. Caffeine: Filtering the facts. New England Journal of Medicine 341:1688–1689.

Etzel, R.A. 1997. Noise: A hazard for the fetus and newborn. Pediatrics 100:724–727.

Groome, L.J., M.J. Swiber, S.B. Holland, L.S. Bentz, J.L. Atterbury, and R.F. Trimm. 1999. Spontaneous motor activity in the perinatal infant before and after birth: Stability in individual differences. Developmental Psychobiology 35:20–24.

Hepper, P.G. 1992. Fetal psychology: An embryonic science. Pp.129–55 in Fetal Behaviour: Developmental and Perinatal Aspects, J.G. Nijhuis, ed. Oxford: Oxford University Press.

Hepper, P.G. 1996. Fetal memory: Does it exist? What does it do? Acta Paediatrica Supplement 416:16–20.

Hepper, P.G., E.A. Shannon, and J.C. Dornan. 1997. Sex differences in fetal mouth movements. Lancet 350:1820.

McCartney, G. and P. Hepper. 1999. Development of lateralized behavior in the human fetus from 12 to 27 weeks' gestation. Developmental Medicine and Child Neurology 41:83–86.

Peiper, A. 1925. Sinnesempfindungen des kindes vor seiner Geburt. Monatsschrift für Kinderheilkunde 29:236–241.

Penn, A.A. and C.J. Schatz. 1999. Brain waves and brain wiring: The role of endogenous and sensory-driven neural activity in development. Pediatric Research 45:447–458.

Teixeira, J.M. and M.N. Fisk. 1999. Association between maternal anxiety in pregnancy and increased uterine artery resistance index: Cohort based study. British Medical Journal 318:153–157.

Weinstock, M. 1997. Does prenatal stress impair coping and regulation of hypothalamic-pituitary-adrenal axis? Neuroscience and Biobehavioral Reviews 21:1–10.

Chapter 2: Birth

Butterworth, G. and B. Hopkins. 1988. Hand-mouth coordination in the newborn baby. British Journal of Developmental Psychology 6:303–314.

DeCasper, A.J. and W.P. Fifer. 1980. Of human bonding. Science 208: 1174–1176.

Dondi, M., F. Simion, and G. Caltran. 1999. Can newborns discriminate between their own cry and the cry of another newborn infant? Developmental Psychology 35:418–426.

Eyer, D.E. 1992. Mother–infant bonding: A scientific fiction. New Haven: Yale University Press.

Fox, N. and R. Davidson. 1986. Taste elicited changes in facial signs of emotion and the asymmetry of brain electrical activity in human newborns. Neuropsychologia 24:417–422.

Haith, M.M. 1986. Sensory and perceptual processes in early infancy. Journal of Pediatrics 109:158–171.

Johnson, M.H., S. Dziurawiec, H.D. Ellis, and J. Morton. 1991. Newborns' preferential tracking of face-like stimuli and its subsequent decline. Cognition 40:1–19.

Kagan, J. 1994. Galen's Prophecy. New York: Basic Books, 204-205.

Karmiloff-Smith, A. 1995. Annotation: The extraordinary cognitive journey from foetus through infancy. Journal of Child Psychology 36:1293–1313.

Kuhl, P. 1994. Learning and representation in speech and language. Current Opinion in Neurobiology 4:812–822.

Lewis, M. 1992. Individual differences in response to stress. Pediatrics 90: 487–490.

Locke, J.L. 1997. A theory of neurolinguistic development. Brain and Language 58:265–326.

Meltzoff, A.N. 1990. Towards a developmental cognitive science. Annals of the New York Academy of Science 608:1–31.

Molfese, D.L., L.M. Burger-Judisch, and L.L. Hans. 1991. Consonant discrimination by newborn infants: Electrophysiological differences. Developmental Neuropsychology 7:177–195.

Molfese, D.L., R.B. Freeman, Jr., and D.S. Palermo. 1975. The ontogeny of brain lateralization for speech and nonspeech stimuli. Brain and Language 2:356–368.

Porter, R.H. and J. Winberg. 1999. Unique salience of maternal breast odors for newborn infants. Neuroscience and Biobehavioral Reviews 23: 439–449.

Rochat, P. 1998. Self-perception and action in infancy. Experimental Brain Research 123:102–109.

Schaal, B., L. Marlier, and R. Soussignan. 1998. Olfactory function in the human fetus: Evidence from selective neonatal responsiveness to odor of amniotic fluid. Behavioral Neuroscience 112:1438–1449.

Slater, A. and R. Kirby. 1998. Innate and learned perceptual abilities in the newborn infant. Experimental Brain Research 123:90–94.

Sugimoto, T., M. Woo, N. Nishida, A. Araki, T. Hara, A. Yasuhara, Y. Kobayashi, and Y. Yamanouchi. 1995. When do brain abnormalities in cerebral palsy occur? An MRI study. Developmental Medical Child Neurology 37:285–292.

Taddio, A., J. Katz, A.L. Ilersich, and G. Koren. 1997. Effect of neonatal circumcision on pain response during subsequent routine vaccination. Lancet 349:599–603.

van der Meer, A.L.H., F.R. van der Weel, and D.N. Lee. 1995. The functional significance of arm movements in neonates. Science 267:693–695.

Chapter 3: Getting Started

Glotzbach, S.F. and D.M. Edgar. 1994. Biological rhythmicity in normal infants during the first 3 months of life. Pediatrics 94:482–488.

Sandyk, R. 1992. Melatonin and maturation of REM sleep. International Journal of Neuroscience 63:105–114.

Shimada, M. and K. Takahashi. 1999. Emerging and entraining patterns of the sleep–wake rhythm in preterm and term infants. Brain and Development 21:468–473.

van den Boom, D.C. 1994. The influence of temperament and mothering on attachment and exploration: An experimental manipulation of sensitive responsiveness among lower-class mothers with irritable infants. Child Development 65:1457–1477.

Wendland-Carro, J. and C.A. Piccinini. 1999. The role of early intervention on enhancing the quality of mother-infant interaction. Child Development 70:713–721.

Worobey, J. and J. Belsky. 1982. Employing the Brazelton scale to influence mothering: An experimental comparison of three strategies. Developmental Psychology 18:736–743.

Chapter 4: Exploring

Bell, M.A. and N.A. Fox. 1994. Brain development over the first year of life. Pp. 314–345 in Human Behavior and the Developing Brain, G. Dawson and K. Fischer, eds. New York: Guilford Press.

Chabris, C.F. 1999. Prelude or requiem for the 'Mozart Effect'? Nature 400:826–827.

Chugani, H.T. 1994. Development of regional brain glucose metabolism. Pp. 153–175 in Human Behavior and the Developing Brain, G. Dawson and K. Fischer, eds. New York: Guilford.

Collie, R. and H. Hayne. 1999. Deferred imitation by 6- and 9-month-old infants: More evidence for declarative memory. Developmental Psychobiology 35:83–90.

Csibra, G., G. Davis, M.W. Spratling, and M.H. Johnson. 2000. Gamma oscillations and object processing in the infant brain. Science 290:1582–1585.

Davidson, R.J. and N.A. Fox. 1989. Frontal brain asymmetry predicts infants' response to maternal separation. Journal of Abnormal Psychology 98:127–131.

Dehaene-Lambertz, G. and S. Dehaene. 1994. Speed and cerebral correlates of syllable discrimination in infants. Nature 370:292–295.

Diamond, A. 1990. The development and neural bases of memory functions as indexed by the AB and delayed response task in human infants and infant monkeys. Annals of the New York Academy of Science 608: 267–303.

Diamond, A. and P.S. Goldman-Rakic. 1989. Comparison of human infants and rhesus monkeys on Piaget's AB task: Evidence for dependence on dorsolateral prefrontal cortex. Experimental Brain Research 74:24–40.

Downs, M.P. and C. Yoshinaga-Itano. 1999. The efficacy of early identification and intervention for children with hearing impairment. Pediatric Clinics of North America 46:79–87.

Fox, N.A., M.A. Bell, and N.A. Jones. 1992. Individual differences in response to stress and cerebral asymmetry. Developmental Neuropsychology 8:161–184.

Gunnar, M.R. 1998. Quality of early care and buffering of neuroendocrine stress reactions: Potential effects on the developing human brain. Preventive Medicine 27:208–211.

Hartshorn, K., C. Rovee-Collier, P. Gerhardstein, R.S. Bhatt, T.L. Wondoloski, P. Klein, J. Gilch, N. Wurtzel, and M. Campos-de-Carvalho. 1998. The ontogeny of long-term memory over the first year-and-a-half of life. Developmental Psychobiology 32:69–89.

Herschkowitz, N., J. Kagan, and K. Zilles. 1997. Neurobiological bases of behavioral development in the first year. Neuropediatrics 28:296–306.

Johnson, M.H. 1994. Brain and cognitive development in infancy. Current Opinion in Neurobiology 4:218–225.

Kinney, H.C., B.A. Brody, A.S. Kloman, and F. Gilles. 1988. Sequence of central nervous system myelination in human infancy. Journal of Neuropathology and Experimental Neurology 47:217–234.

Kuhl, P.K. and A.N. Meltzoff. 1982. The bimodal perception of speech in infancy. Science 218:1138-1140.

Mandler, J.M. and L. McDonough. 1993. Concept formation in infancy. Cognitive Development 8:291–318.

Quinn, P.C. and P.D. Eimas. 1993. Evidence for representations of perceptually similar natural categories by 3-month-old and 4-month-old infants. Perception 22:463–475.

Sininger, Y.S., K.J. Doyle, and J.K. Moore. 1999. The case for early identification of hearing loss in children. Pediatric Clinics of North America 46: 1–14.

Younger, B.A. and D.D. Fearing. 1999. Parsing items into separate categories: Developmental change in infant categorization. Child Development 70:291–303.

Zentner, M.R. and J. Kagan. 1996. Perception of music by infants. Nature 29:383.

Chapter 5: Comfort and Communication

Child, L.M.F. 1992. The Mother's Book. Bedford, Mass.: Applewood Books, 1.

Davidson, R. and K. Hugdahl, eds. 1998. Brain Asymmetry. Cambridge: MIT Press.

Dunn, J. The beginnings of moral understanding: Development in the second year. Chapter 2 in The Emergence of Morality in Young Children, J. Kagan and S. Lamb, eds. Chicago: University of Chicago Press.

Greenspan, S.I. 1991. Clinical assessment of emotional milestones in infancy and early childhood. Pediatric Clinics of North America 38:1371–1386.

Gunnar, M.R., L. Brodersen, K. Krueger, and J. Rigatuso. 1996. Dampening of adrenocortical responses during infancy: Normative changes and individual differences. Child Development 67:877–879.

Hewlett, B.S. and M.E. Lamb. 1998. Culture and early infancy among central African foragers and farmers. Developmental Psychology 34:653–661.

Locke, J.L. 1990. Structure and stimulation in the ontogeny of spoken language. Developmental Psychobiology 23:621–643.

MacWhinney, B. 1998. Models of the emergence of language. Annual Review of Psychology 49:199–202.

NICHD, Early Child Care Research Network. 1999. Child care and mother-child interaction in the first 3 years of life. Developmental Psychology 35:1399–1413.

Sroufe, L.A. and E. Waters. 1976. The ontogenesis of smiling and laughter: A perspective on the organization of development in infancy. Psychological Reviews 83:173–189.

Tronick, E.Z. and J.F. Cohn. 1989. Infant–mother face-to-face interaction: Age and gender differences in coordination and the occurrence of miscoordination. Child Development 60:85–92.

van den Boom, D.C. 1997. Sensitivity and attachment: Next steps for developmentalists. Child Development 64n.4:592–593.

Chapter 6: Discovering

Bloom, L., C. Margulis, E. Tinker, and N. Fujita. 1996. Early conversations and word learning: Contributions from child and adult. Child Development 67:3154–3175.

Bornstein, M.H., C.S. Tamis-LeMonda, and O.M. Haynes. 1999. First words in the second year: Continuity, stability, and models of concurrent and predictive correspondence in vocabulary and verbal responsiveness across age and context. Infant Behavior & Development 22n.1:65–85.

Herschkowitz, N., J. Kagan, and K. Zilles. 1999. Neurobiological bases of behavioral development in the second year. Neuropediatrics 30:221–230.

Huttenlocher, J. 1998. Language input and language growth. Preventive Medicine 27:195–199.

Landry, S.H., K.E. Smith, P.R. Swank, and C.L. Miller-Loncar. 2000. Early maternal and child influences on children's later independent cognitive and social functioning. Child Development 71:358–375.

Lucariello, J. 1987. Concept formation and its relation to word learning and use in the second year. Journal of Child Language 14:309–332.

McCarty, M.E., R.R. Collard, and R.K. Clifton. 1999. Problem solving in infancy: The emergence of an action plan. Developmental Psychology 35:1091–1101.

Molfese, D.L. 1989. Electrophysiological correlates of word meanings in 14-month-old human infants. Developmental Neuropsychology 5:79–103.

Molfese, D.L. 1990. Auditory evoked responses recorded from 16-month-old human infants to words they did and did not know. Brain and Language 38:345–363.

Chapter 7: Me and You

Bates, E. 1990. Language about me and you: Pronominal reference and the emerging concept of self. Pp.1–5 in The Self in Transition: Infancy to Childhood, D. Ciccetti and M. Beeghly, eds. Chicago: University of Chicago Press.

Bauer, P.J. and G.A. Dow. 1994. Episodic memory in 16- and 20-month-old children: Specifics are generalized but not forgotten. Developmental Psychology 30:403–417.

Carpenter, M., N. Akhtar, and M. Tomasello. 1998. Fourteen through eighteen-month-old infants differentially imitate intentional and accidental actions. Infant Behavior and Development 21:315–330.

Case, R. 1992. The role of the frontal lobes in the regulation of cognitive development. Brain and Cognition 20:51–73.

Chandler, M., A.S. Fritz, and S. Hala. 1989. Small-scale deceit: Deception as a marker of two-, three-, and four-year-olds' early theories of mind. Child Development 60:1263–1277.

Damon, W. 1988. Empathy, shame, guilt. Pp.13–29 in The Moral Child. New York: Free Press.

Davidson, R.J. 1994. Temperament, affective style and frontal lobe asymmetry. Pp.518–537 in Human Behavior and the Developing Brain, G. Dawson and K. Fischer, eds. New York: Guilford Press.

Howe, M.L. and M.L. Courage. 1997. The emergence and early development of autobiographical memory. Psychological Review 104:499–523.

Kochanska, G., R.J. Casey, and A. Fukumoto. 1995. Toddlers' sensitivity to standard violations. Child Development 66:643–656.

Kochanska, G., D.R. Forman, and K.C. Coy. 1999. Implications of the mother–child relationship in infancy for socialization in the second year of life. Infant Behavior & Development 22:249–265.

Lewis, M. 1995. Self-conscious emotions. American Scientist 83:68–78.

Meltzoff, A. 1995. Understanding the intentions of others: Re-enactment of intended acts by 18-month-old children. Developmental Psychology 31:838–850.

Repacholi, B.M. and A. Gopnik. 1997. Early reasoning about desires: Evidence from 14- and 18-month-olds. Developmental Psychology 33:12–21.

Rothbart, M.K. and S.A. Ahadi. 1994. Temperament and the development of personality. Journal of Abnormal Psychology 103:55–66.

Zahn-Waxler, C., M. Radke-Yarrow, and E. Wagner. 1992. Development of concern for others. Developmental Psychology 28:126–36.

Chapter 8: Gaining Competence

Berk, L.E. 1994. Why children talk to themselves. Scientific American November:60–65.

Csikszentmihalyi, M. 1996. Creativity. New York: Harper Perennial.

Damasio, A.R., D. Tranel, and H. Damasio. 1990. Individuals with sociopathic behavior caused by frontal damage fail to respond autonomically to social stimuli. Behavioral Brain Research 41:81–94.

Fischer, P.M., M.P. Schwarz, J.W. Richards, Jr., A.O. Goldstein, and T.H. Rojas. 1991. Brand logo recognition by children aged 3 to 6 years. Journal of the American Medical Association 266:3145–3148.

Gardner, W. and B. Rogoff. 1990. Children's deliberateness of planning according to task circumstances. Developmental Psychology 26:480–487.

Gathercole, S.E. 1998. The development of memory. Journal of Child Psychology and Psychiatry 39:3–27.

Giedd, J.N., J.M. Rumsey, F.X. Castellanos, J.C. Rajapakse, D. Kaysen, A.C. Vaituzis, Y.C. Vauss, S.D. Hamburger, and J.L. Rapoport. 1996. A quantitative MRI study of the corpus callosum in children and adolescents. Brain Research Developmental Brain Research 91:274–280.

Goldman-Rakic, P.S. 1992. Working memory and the mind. Scientific American 267n.3:111–117.

Grattan, L.M. and P.J. Eslinger. 1991. Frontal lobe damage in children and adults. Developmental Neuropsychology 7:283–286.

Grattan, L.M. and P.J. Eslinger. 1992. Long-term psychological consequences of childhood frontal lobe lesion in patient DT. Brain and Cognition 20:185–195.

Harris, J.C. 1995. Emergence of the self. Pp.219–233 in Developmental Neuropsychiatry, Vol. 1. Oxford: Oxford University Press.

Hayne, H., S. MacDonald, and R. Barr. 1997. Developmental changes in the specificity of memory over the second year of life. Infant behavior and development 20:233–245.

Howe, M.L. and M.L. Courage. 1997. The emergence and early development of autobiographical memory. Psychological Review 104:499–523.

Murphy, B.L., A.F.T. Arnsten, P.S. Goldman-Rakic, and R.H. Roth. 1996. Increased dopamine turnover in the prefrontal cortex impairs spatial working memory performance. Proceedings of the U.S. National Academy of Science 1325–1329.

Nelson, C.A. 1998. The nature of early memory. Preventive Medicine 27: 172–179.

Siegler, R.S. 2000. The rebirth of children's learning. Child Development 71:26–35.

Singer, W. 1993. Synchronization of cortical activity and its putative role in information processing and learning. Annual Review of Physiology 56:349–374.

Vurpillot, E. 1968. The development of scanning strategies and their relation to visual differentiation. Journal of Experimental Child Psychology 6:632–650.

Winsler, A., R.M. Diaz, and I. Montero. 1997. The role of private speech in the transition from collaborative to independent task performance in young children. Early Childhood Research Quarterly 12:59–79.

Zelazo, P.D., D. Frye, and T. Rapus. 1996. An age-related dissociation between knowing Rules and using Them. Cognitive Development 11: 37–63.

Chapter 9: Living Together

Cassidy, K.W., J.Y. Chu, and K.K Dahlsgaard. 1997. Preschoolers' ability to adopt justice and care orientations to moral dilemmas. Early Education & Development 8:419–434.

Coyle, J.T. 2000. Psychotropic drug use in very young children. Journal of the American Medical Association 283:1059–1060.

Hoffner, C. and J. Cantor. 1985. Developmental differences in response to a television character's appearance and behavior. Developmental Psychology 21:1065–1074.

Kagan, J. 1999. The role of parents in children's psychological development. Pediatrics 104:164–167.

Shoda, Y., W. Mischel, and P.K. Peake. 1990. Predicting adolescent cognitive and self-regulatory competencies from preschool delay of gratification: Identfying diagnostic conditions. Developmental Psychology 26:978–986.

Tremblay, R., R.O. Pihl, F. Vitaro, and P. Dobkin. 1994. Predicting early onset of male antisocial behavior from preschool behavior. Archives of General Psychiatry 51:732–739.

Zito, J.M., D.J. Safer, S. dosReis, J.J. Gardner, M. Boles, and F. Lynch. 2000. Trends in the prescribing of psychotropic medications to preschoolers. Journal of the American Medical Association 283:1025–1030.

Chapter 10: Paths to Personality

Boyce, W.T., R.G. Barr, and L.K. Zeltzer. 1992. Temperament and the psychobiology of childhood stress. Pediatrics 90:483–486.

Cloninger, C.R. and R. Adolfsson. 1996. Mapping genes for human personality. Nature Genetics 12:3–4.

Davidson, R.J. 1992. Anterior cerebral asymmetry and the nature of emotion. Brain and Cognition 20:125–151.

DiLalla, L.F., J. Kagan, and J.S. Reznick. 1994. Genetic etiology of behavioral inhibition among 2-year-old children. Infant Behavior and Development 17:401–408.

Fox, N.A., L.A. Schmidt, S.D. Calkins, K.H. Rubin, and R.J. Coplan. 1996. The role of frontal activation in the regulation and dysregulation of social behavior during preschool years. Development and Psychopathology 8:89–102.

Gunnar, M.R., M. deHaan, S. Pierce, K. Stansbury, and K. Tout. 1997. Temperament, social competence, and adrenocortical activity in preschoolers. Developmental Psychobiology 31:65–85.

Kagan, J., M. Julia-Sellers, and M.O. Johnson. 1991. Temperament and allergic symptoms. Psychosomatic Medicine 53n.3:332–340.

Kagan, J., N. Snidman, M. Zentner, and E. Peterson. 1999. Infant temperament and anxious symptoms in school age children. Development and Psychopathology 11:209–224.

Kochanska, G. 1997. Multiple pathways to conscience for children with different temperaments: From toddlerhood to age 5. Developmental Psychology 33:228–240.

Lamb, M.E. and M.H. Bornstein. 1987. Development in Infancy: An Introduction. New York: Random House, 236.

Lemonick, M.D. 1999. Smart genes. Time September 13:52–59.

Lewis, M. 1992. Individual differences in response to stress. Pediatrics 90: 487–490.

Plomin, R. and J.C. DeFries. 1998. The genetics of cognitive abilities and disabilities. Scientific American May:40–47.

Sternberg, R.J. 2000. The Holey Grail of general intelligence. Science 289: 399–401.

Index

Italicized page numbers refer to illustrations.